AF811533

Design Science and Innovation

Series Editor

Amaresh Chakrabarti, Centre for Product Design and Manufacturing,
Indian Institute of Science, Bengaluru, India

The book series is intended to provide a platform for disseminating knowledge in all areas of design science and innovation, and is intended for all stakeholders in design and innovation, e.g. educators, researchers, practitioners, policy makers and students of design and innovation. With leading international experts as members of its editorial board, the series aims to disseminate knowledge that combines academic rigour and practical relevance in this area of crucial importance to the society.

Balkrishna C. Rao

Frugal Engineering

Advent, Design and Production of Frugal Products

Springer

Balkrishna C. Rao
Department of Engineering Design
Indian Institute of Technology Madras
Chennai, India

ISSN 2509-5986 ISSN 2509-5994 (electronic)
Design Science and Innovation
ISBN 978-981-99-9699-5 ISBN 978-981-99-9700-8 (eBook)
https://doi.org/10.1007/978-981-99-9700-8

© The Editor(s) (if applicable) and The Author(s), under exclusive license to Springer Nature Singapore Pte Ltd. 2024

This work is subject to copyright. All rights are solely and exclusively licensed by the Publisher, whether the whole or part of the material is concerned, specifically the rights of translation, reprinting, reuse of illustrations, recitation, broadcasting, reproduction on microfilms or in any other physical way, and transmission or information storage and retrieval, electronic adaptation, computer software, or by similar or dissimilar methodology now known or hereafter developed.
The use of general descriptive names, registered names, trademarks, service marks, etc. in this publication does not imply, even in the absence of a specific statement, that such names are exempt from the relevant protective laws and regulations and therefore free for general use.
The publisher, the authors and the editors are safe to assume that the advice and information in this book are believed to be true and accurate at the date of publication. Neither the publisher nor the authors or the editors give a warranty, expressed or implied, with respect to the material contained herein or for any errors or omissions that may have been made. The publisher remains neutral with regard to jurisdictional claims in published maps and institutional affiliations.

This Springer imprint is published by the registered company Springer Nature Singapore Pte Ltd.
The registered company address is: 152 Beach Road, #21-01/04 Gateway East, Singapore 189721, Singapore

Paper in this product is recyclable.

Dedicated to my mother
Usha Rao

वेन्कटाद्री समम् स्थानम् ब्रह्माण्डे नास्ति किञ्चन
वेन्कटेश समादेवो न भूतो न भविष्यति

(The sacred hills of Venkatadri are unique in this universe
Shri Venkateshwara, residing on these hills, is the divine Lord
who reigns supreme forever)

Preface

This book is the first edition on frugal engineering, an emerging field that has received a lot of attention mainly due to the planetary and human-made crises facing this planet. Technological progress picked up only in the last couple of centuries and nations the world over never went *frugal* at that time when resources were relatively abundant and climate change was gradually rearing its head. Their early foray into frugality could have made it a significant influencer in all engineering endeavors and might have slowed the pace of climate change and subsequently helped in the ensuing mitigation efforts. Notwithstanding this delay, it is noteworthy that the world has started taking frugality seriously over the past couple of decades.

Against the above backdrop, this book aims to present a change that is urgently warranted for engineering activities in this era of widespread sustainable development. Climate change and other crises are beckoning us to restrain our use of resources while tempering our consumption of energy. The concept of being frugal gels very well with the current need to use resources judiciously while uplifting the living standards of an increasing global population. The first edition of this book presents the philosophy and related concepts of frugal engineering. Although frugal engineering is based on a new amalgamation of classical and modern scientific principles (as every concept is!), the use of frugality actually leads in many cases to novel versions of products/services not hitherto seen in the practice of traditional engineering. This book focuses on the development of frugal products but the concepts developed here can also be applied for the development of frugal services possibly using frugal products. The word "frugal" is an umbrella term encompassing all activities and products existing and/or created to tackle emissions, resources and costs while maintaining the best functionality. It should be noted that this book was majorly written during the pandemic and as such has references to it in many sections and chapters. A hopeful future edition will change this tone.

I am greatly indebted to the Lord of Tirumala, Sriman Venkateshwara and Mother Padmavati Akka Devi for their divine guidance whose absence would not have made this book possible! There are several people to thank who in some way or another have contributed immensely to the creation of this book. Foremost, I would like to thank my parents Ramachandra Rao and Usha Rao; brothers Sanjay and Shrikanth; sister-in-law Suchitra; and nephews Atul and Rohit for their significant help in providing a rich family atmosphere rooted in good values. I am thankful to Professor Srinivasan Chandrasekar of Purdue University, USA, whose guidance, wealth of knowledge and kind help have helped me in myriad ways in research and teaching. I am thankful to a whole bunch of my colleagues at Purdue University, IIT Madras and IISc Bengaluru for their camaraderie. A perk of academic life is working with talented research students and I am very thankful for the rich associations with talented students I have worked with until now. I am thankful to Ms. Muskan Jaiswal, Ms. Swati Meherishi and their teams at Springer for their support in making this book possible.

Chennai, India
September 2023

Prof. Balkrishna C. Rao
ICCR Visiting Chair (Jun–Oct, 2022),
Leibniz University Hannover

Contents

About the Author

Balkrishna C. Rao is Professor in the Department of Engineering Design at the Indian Institute of Technology (IIT) Madras. Prof. Rao has served from June 2022 to October 2022 as the ICCR Visiting Chair Professor of Indian Studies in Leibniz University Hannover, where he taught and conducted research on frugal engineering of advanced frugal products. His research interests include sustainable manufacturing and frugal engineering. In frugal engineering, he has coined the term "advanced frugal innovation (AFI)" and also developed the factor of frugality approach to designing and fabricating advanced frugal products in any sector. He has also developed the concept of net finished shape (NFS) for zero waste in frugal manufacturing. Prof. Rao earned his Ph.D. in Mechanical Engineering from Purdue University, US. He has also served as a research associate at Purdue with the School of Industrial Engineering. The long stint at Purdue has resulted in research contributions to the automotive- and aerospace-sectors in the US with involvement in nanotechnology for manufacturing. At IIT Madras, Prof. Rao uses his long research experience in engineering and manufacturing for infusing frugality into product development for a better world.

Chapter 1
Advent of Frugal Innovations

Frugal innovations are all the rage (Tiwari and Herstatt 2012; Rao 2013; Knorringa et al. 2016; Prabhu 2017; Miesler et al. 2020). Their advent is mainly attributed to India, followed by China and, subsequently found their way into Western management principles (Economist 2010). These innovations have come a long way since their beginning at the humble grassroots levels of society. The inherent natural scarcity of resources at the bottom of the pyramid (Rao 2018) lead to the creation of clever contraptions for satisfying the needs of the people involved. Critical to this development are the presence of clever and inventive individuals whose ideas significantly aided in innovating out of the shortages hindering the daily lives of the low-income group. An interesting and very useful grassroots frugal product created in and made for the low-income segment is *MittiCool* (Raturi 2010; Rao 2013), a refrigerator made out of earthenware that requires no artificial power source for its working. More details about this innovation are given in the chapter on *Grassroots Frugal Innovations*.

Moreover, the principles of frugality developed for grassroots products have percolated into sophisticated products. In other words, frugal products are being systematically engineered from scratch for applications involving the society at large. However, the increasing sophistication in going from *grassroots* to *advanced frugal* products (Rao 2017) requires meticulous application of science, technology and cutting-edge research. So the design and fabrication of *frugal innovations* is not mundane and actually uses scientific progress to its advantage. In fact, advances are being made in science and technology through frugal endeavors and the world is learning to do top quality science with frugal resources (Rao 2017).

It is important to note that the *frugal innovation* movement started with innovating in settings with scarcer resources. This is in stark contrast with innovating in settings where ample resources are provided as in the countries of the *organization of economic cooperation and development* (OECD) and also richer segments of emerging markets (Rao 2019). It should be noted that the term resources is broadly applied here to include knowledge, finance, personnel and any other commodity required for

© The Author(s), under exclusive license to Springer Nature Singapore Pte Ltd. 2024
B. C. Rao, *Frugal Engineering*, Design Science and Innovation,
https://doi.org/10.1007/978-981-99-9700-8_1

successfully carrying out the innovation process. Although innovation under *frugal* conditions is low-cost or cheap, the idea underlying the innovation in most cases is brilliant. In fact, many of these ideas have been frugally setup in settings of OECD countries by companies such as General Electric (GE) (Rao 2013). The portable *electrocardiogram* and *ultrasound* developed by GE's healthcare division have been successfully applied in the US and Europe after their initial foray into India and China (Rao 2013).

Over the years, frugality has seeped into Western countries of the world including many parts of Europe and the US to cater to the needs of their populations. In doing so, frugal innovations of the *advanced* type have entered various strata of this society. These innovations aid emergency situations, affecting both the rich and poor, by offering products that are portable and with a performance at par or even better than the traditional clunky products. Other than emergency situations, frugal innovations have also aided in the maintenance and upliftment of living standards of the low-income group of people in the developed world. Such low-income group includes both the poor and the swelling ranks of unemployed workers resulting from human-made crises such as the 2007–09 financial crisis in the US (World-Bank 2009; Rao 2013).

Also, the pandemic related to Covid-19 has shown the significant potential *frugal innovations* possess in responding to emergency situations (Corsini et al. 2021). Several *frugal* solutions for simple face-masks and sophisticated ventilators have shown the significant leveraging of these innovations that is possible provided they are implemented in a planned and meticulous manner.

Against this background there is a need for a book that clearly delineates the concepts and principles underlying frugal innovations and also highlights interesting applications relevant to this era of sustainability. Accordingly, this book is an introduction to the interesting world of *frugal engineering* for abetting widespread sustainable development, the world over. This book is an invitation to one and all, including workers with advanced skills, to the interesting world of *frugal engineering* and also a pointer to the significant difference that meticulous involvement of science, engineering, economics and other knowledge-bases can make to these innovations. Furthermore, the planetary and human-made crises, with the financial mess of 2007–09 being an instance of the latter, of our time are beckoning us to join the *frugal* bandwagon. In this regard, the clubbing of *classical* or traditional principles and also newer methodologies under the *frugal* banner will hopefully encourage and empower many of us into going green.

It should be noted that subsequent chapters of this book have referred to applications from various sectors in various entities of different countries as *Advanced Frugal Innovations*. This terminology is based only on author's research efforts and does not represent the opinion of their manufacturers. In other words, Chap. 4 cites many examples of such innovations but they have not been classified as *Advanced Frugal* by their creators/manufacturers. The author has combed several sources and presented the resulting findings in Chap. 4 as representatives of *Advanced Frugal Innovations*.

This book is divided into ten chapters with subsequent chapters on the following:
Chapter 2: Frugal Innovations and Sustainable Development
Chapter 3: Grassroots Frugal Innovations
Chapter 4: Advanced Frugal Innovations
Chapter 5: Engineering of Frugal Products
Chapter 6: Frugal Design Approach
Chapter 7: Complex Systems and Frugal Products
Chapter 8: Frugal Manufacturing
Chapter 9: Quality Control of Frugal Products
Chapter 10: Combating the COVID-19 Pandemic with Advanced Frugal Innovations
Chapter 11: Frugal Computing for Artificial Intelligence and Other Applications
Chapter 12: Combating Poverty with Frugal Innovations
Chapter 13: The Future.

1.1 India's Proud Heritage in Science and Mathematics

India has been documented for its strong heritage encompassing advances in metallurgy, medicine, mathematics and astronomy over millennia BC. In fact, the *Kerala* school of mathematics has been acknowledged for their contribution to calculus and that concepts underlying numerical integration, infinite-series for π, and elementary functions such as $\sin x$, $\cos x$, and $\tan^{-1} x$ were known to *Kerala*-mathematicians over 250 years earlier than those documented by *Newton* and *Leibniz* (Joseph 2016). *Madhava* of *Sangamagrama* is a famous representative of the *Kerala* school of mathematics (Joseph 2016). Indian metallurgists of the early AD era had devised downward distillation of vaporized zinc by smelting zinc ore through special retorts comprising condensers and furnaces to obtain melt that subsequently gave solid zinc (Srinivasan and Ranganathan 2024). *Nagarjuna* in *Rasaratnakara* describes this ingenious method of producing zinc when the world did not know of one (Srinivasan and Ranganathan 2024)! Another achievement in materials science was the Delhi iron pillar. This pillar located in Delhi was built in the third century AD during the golden period of the *guptas* (Srinivasan and Ranganathan 2024). The pillar, shown in Fig. 1.1, was manufactured through the brilliant technique of forge-welding and it has shown no signs of rusting!. Other than being the metallurgical wonder of that time, the pillar was also a marvel of astronomy wherein its shadow was used to fix the arrival of the summer solstice (Balasubramaniam and Prabhakar 2007). In fact, India was also popular during the start of the Christian era for its *Wootz* steel, a high carbon iron alloy, made from the crucible process. The *Wootz* steel which was exported to Europe, China, and the Arab block was also believed to be a crucial ingredient of the awe inspiring *Damascus* blade (Srinivasan and Ranganathan 2024; Srinivasan and Griffiths 1997), see Fig. 1.2 (Sherby and Wadsworth 2001).

The brief diversion into a glimmer of India's rich heritage clearly brings out India's strong penchant for ingeniously tackling challenging problems in science and mathematics. Such talent, especially for tackling tough abstract problems, has immensely

Fig. 1.1 The iron pillar of Delhi. A testimony to India's rich tradition in mathematics and science. Image taken from Balasubramaniam and Prabhakar (2007)

helped India grow and prosper over the years. And this inventive streak also gave the concept of *frugal innovations* to the world. This book with its strong emphasis on advances in science for frugality strongly believes that there is a strong Indian undercurrent to the brilliant *advanced frugal innovations* described here and elsewhere. The reader is entreated to focus on the complexity lurking beneath the simplicity of these innovations that need a strong scientific bent of mind to conceptualize and realize. This book is an introduction that will aid beginners and experts in their foray into the wonderful world of *advanced frugal innovations*.

1.2 "Advanced Frugal" is NOT Jugaad!!

This oddly named final section of the chapter has been added to clarify the importance of going *frugal* in this exciting age. Although, the past couple of decades has seen *climate change* gradually influencing all human endeavors, it has also seen a significant burst of new development in science and engineering that are attempting to make our planet better. Such developments have to keep a keen focus on our plan-

Fig. 1.2 Damascus swords known for their sharp and effective cutting edge are strongly believed to be made of *Wootz* steel from India. Image taken from Sherby and Wadsworth (2001)

etary crises while creating newer and better products and services. In this regard, the *advanced frugal innovations or products* of this book bring out the immense engineering challenges that have been surmounted using scientific principles and scientific tools. In fact, many of the examples of this book have used cutting-edge research to create these products while strictly constraining the resources going into their making.

The astute reader will not see the hand of *Jugaad* in the myriads of examples showcased in this book. This is not to belittle *Jugaad* whose features of low-cost and lesser resources have been instrumental in arriving at some of the concepts of this book. Hence, *frugal engineering* of *advanced frugal products* do NOT involve any makeshift collection of parts for their working. However, such *advanced frugal* products need the Indian ingenuity while applying cutting-edge research in their creation. We do not have multiple earths to squander our resources wantonly. And *frugal engineering* has arrived apropos beckoning the world to take important cues from India in applying cutting-edge science and other knowledge areas for tackling the planetary and human-made crises of our time. *Advanced frugal* products need every bit of the best science and engineering has to offer for SYSTEMATICALLY building such products from scratch. This is borne out by the many examples of *advanced frugal products* from several sectors presented in this book whose sophisticated design and working attest to the brilliant use of science and engineering.

1.3 Why This Book?

This book is an introduction to *frugal engineering* of any product in any sector. It also covers the interesting world of *advanced frugal products* that result after the *frugal engineering* of a given product. However, this book also brings out the urgent need for change in engineering itself for handling *climate change*. The conservative designs coming out of current engineering practice, that we take for granted, are not feasible in the long run. This is because conservative practices in engineering activities consume resources in excess of what is usually needed for an effective design. And excess consumption of resources typically translates into higher energy usage and hence higher attendant emissions of *green house gases*. Therefore, it is imperative from here onwards to gear engineering activities for producing streamlined products which consume minimal resources for their creation, operation and disassembly. And such streamlined products' functionality should also be affordable to society at large. This is necessary for mass upliftment of living standards that also helps in pulling people out of penury. Therefore, engineering and its practitioners, i.e., engineers and scientists, bear the utmost responsibility of building products and services that tackle *climate change* and uplift living standards!

References

Balasubramaniam R, Prabhakar VN (2007) On technical analysis of characters of the oldest Delhi iron pillar inscription. Curr Sci 92(12):1709–1719

Corsini L, Dammicco V, Moultrie J (2021) Frugal innovation in a crisis: the digital fabrication maker response to covid-19. R&D Manag 51:195–210

Economist (17 Apr 2010) First break all the rules: the charms of frugal innovation

Joseph G (2016) Indian mathematics: engaging with the world from ancient to modern times. World Scientific, New Jersey

Knorringa P, Pesa I, Leliveld A, van Beers C (2016) Frugal innovation and development: aides or adversaries? Eur J Dev Res 28(2):143–153

Miesler T, Wimschneider C, Brem A, Meinel L (2020) Frugal innovation for point-of-care diagnostics controlling outbreaks and epidemics. ACS Biomater Sci Eng 6(5):2709–2725

Prabhu J (2017) Frugal innovation: doing more with less for more. Philos Trans R Soc A: Math Phys Eng Sci A375:20160372

Rao B (2013) How disruptive is frugal? Technol Soc 35:65–73

Rao B (2017) Advances in science and technology through frugality. IEEE Eng Manag Rev 45(1):32–38

Rao B (2018) Science is indispensable to frugal innovations. Technol Innov Manag Rev (TIM) 8(4):49–56

Rao B (2019) The science underlying frugal innovations should not be frugal. R Soc Open Sci 6(5)

Raturi P (2010) Potter power: son of the soil. Entrepreneur, pp 108–109

Sherby OD, Wadsworth J (2001) Ancient blacksmiths, the Iron Age, Damascus steels, and modern metallurgy. J Mater Process Technol 117:347–353

Srinivasan S, Griffiths D (1997) Crucible steel from south India, preliminary investigations on some newly identified sites, vol 462. Materials Research Society, Pittsburgh, USA

Srinivasan S, Ranganathan S (n.d.) Metallurgical heritage of India. https://www.tf.uni-kiel.de/matwis/amat/def_en/articles/metallurg_heritage_india/metallurgical_heritage_india.html

Tiwari R, Herstatt C (2012) Open global innovation networks as enablers of frugal innovation: propositions based on evidence from India. Hamburg, Germany. In: Institute for Technology and Innovation Management Working Paper, Hamburg, Germany

World-Bank (2009) The global economic crisis: assessing vulnerability with a poverty lens, policy note. World Bank, Washington, DC

Chapter 2
Frugal Innovatons and Sustainable Development

The dawning of the 21st century has been accompanied by natural and human-made-crises along with the forces of globalization. First, independent institutions such as the Intergovernmental Panel on Climate Change (IPCC) have identified climate change as a serious threat to our way of life (IPCC 2001; IEA 2009; MEA 2005; Rao 2013), which is supplemented by other threats including losses-in-bio-diversity and scarcer resources (Rockström et al. 2009). At the very least, these natural crises beg for a spartan lifestyle to preserve our complex ecosystems for the betterment of current and future generations. Human-made threats such as the dot-com bubble (Lowenstein 2004) and the recent financial crises (World-Bank 2009) in particular, are also prodding the global populace toward thrift. Second, globalization is expanding the ranks of the working class in emerging economies, whose desire for higher standards of living (Bordo et al. 2002; World-Bank 2002) is putting a strain on our resources. It should be noted that despite reports on the on-going **covid-19** pandemic battering its forces, *Globalization* is here to stay due to its benefits over the years to all the stakeholders including endusers involved in the game. Moreover, perturbations due to the above mentioned threats will get aggravated due to the forecasted explosion in world population (UN-Population-Division 1999).

However, before proceeding any further, a note on *climate change*, since this book believes this menace to be the defining problem of our time, which is affecting and also amplifying other crises alluded to in the above paragraph. Earth has been exposed to periodic bouts of warming in the deep past based on the geological timescale. However, such warming typically resulted from major volcanic eruptions that engulfed a great portion of this planet. In contrast, the current spell of warming is influenced by humans due to our energy guzzling and hence emission belching activities involving *green house gases* (GHGs) that block outgoing solar radiation which subsequently warms up the Earth's surface. Although these endeavors were mainly agrarian since humans started settlements, various industrial activities have been complementing our agricultural work since the dawn of the industrial revolution (MacKay 2009). In fact, even agriculture itself has gone through mechanization over

© The Author(s), under exclusive license to Springer Nature Singapore Pte Ltd. 2024
B. C. Rao, *Frugal Engineering*, Design Science and Innovation,
https://doi.org/10.1007/978-981-99-9700-8_2

the past centuries thereby being a party to *climate change*. Overall, the phemonenon of *climate change* is a significant problem that is also interacting and aggrevating other crises including declines in *biodiversity* and natural resources. Conversely, losses in ecosystems and resources and population-explosion are also positively feeding into *climate change* and amplifying it through emissions of GHGs other than carbon (Nature-Editorial 2021). Therefore, human induced planetary crises have altered Earth with geologists even seriously mulling to call this era the *Anthropocene* (Steffen et al. 2011; Hamilton 2016) for signifying the "human" touch. Therefore, it behoves us to take these crises seriously and use our intellectual and collective capabilities for the betterment of Earth against their onslaught.

2.1 Frugality an Important Tool of Sustainability

The concept of sustainability borrowed from forestry studies was formally established for the wider world in 1987 by the *Brundtland Commission*. The term *sustainable development* (SD) was coined in 1987 by the *United Nations World Commission on Environment and Development* and defined as "development that meets the needs of the present without compromising the ability of future generations to meet their own needs" (Brundtland-Commission 1987).

The continued progression of both *climate change* and other planetary crises has made it indispensable to leave a minimal carbon footprint in all activities of every nation on this planet. The prospect of the global population turning abstemious, against the above backdrop, is dim due to the aspirations of an expanding middle class in emerging markets and also rich nations shrugging to opt for simpler lifestyles. Moreover, planetary and man- made crises, such as the recent financial meltdown of 2008–2009 (World-Bank 2009), are encouraging the development of low-cost products and low-cost services that are affordable to society at large for improving their living standards (Rao 2014). The importance of sustainability and low price is borne out by current trends. On the one hand, natural threats are empowering governments world over to enact stringent regulations on emissions and energy efficiency. This is exemplified by the adoption of carbon trading schemes and/or energy efficient technologies and/or emissions limits by various economies of the developed world and the BRICs (Brazil, Russia, India and China) (Barbier 2010; Ministry of Finance, India 2010; Nature 2010). On the other hand, cost-conscious consumers in emerging markets are demanding quality products and services at a discount for better lifestyles.

An appealing solution to this dilemma is the use of frugal innovations with cutting-edge technology to create low-cost and no-frills products. Therefore, the twin challenges of affordability and sustainability can be tackled by the widespread usage of frugal products and frugal services. This is because, other than their lower costs, frugal innovations also possess green credentials due to their no-frills nature that entails lesser consumption of resources and hence generally lesser emissions (Moynihan and Allwood 2014), which are both beneficial to the environment. Hence, companies will have to adopt the principles of frugal innovations to achieve profitability against both cost-conscious consumers and regulations for sustainable development.

The adoption of frugality entails design principles that advocate minimal use of resources for realizing efficient functioning of products. Other than minimal amounts of raw materials, the economizing of resources would also involve the reuse of components and simpler designs that result in products without extra accessories. The no-frills nature of frugal products makes them "lean" when compared to their "ordinary" counter-parts, which are bulky due to consumption of relatively large amounts of resources. Consequently, aiming for a no-frills product helps maintain lower costs with a positive impact on sustainability because of lesser resource consumption.

By economizing the resource consumption and aiming for simple products and services, albeit sometimes using cutting-edge technology, frugal innovations have a lot to offer for sustainable development. As a result, rich countries, such as the U.S., whose technological growth in recent years has "flattened (Hamilton 2016) should join the frugal bandwagon to give a boost to technological growth that will impact both economic growth and employment. Even export powerhouses like Germany with its legendary Mittelstand will have to embrace these innovations to cater to cost- and environment-conscious customers of the future".

2.2 Frugality Can Abet Adoption of Sustainability

Although frugal innovation is characterized as a grassroots phenomenon that caters mainly to the bottom-of-the-pyramid, the all-inclusive applications of their sophisticated brethren are globally uplifting both the rich and poor. Some instances of this inclusivity include portable versions of *electrocardiogram* (ECG) and *ultrasound* developed by *General Electric* (GE), which have found wide spread applications in the society at large (Immelt et al. 2009). The appearance of frugality in a range of advanced applications cannot be neglected and is a force for good as far as sustainable development is concerned. The conflicting goals of achieving better standards of living at lower costs while having a minimal impact on the environment demand ever more rigorous involvement of scientific research in frugality, which is important for sustainable development.

References

Barbier E (2010) A global green new deal: rethinking the economic recovery. Cambridge University Press, New York, USA

Bordo M, Taylor A, Williamson J (2003) Globalization in historical perspective. The University of Chicago Press, USA

Brundtland-Commission (1987) Our common future. Technical report, UN World Commission on Environment and Development

Hamilton C (2016) World view: Define the anthropocene in terms of the whole earth. Nature 536:251

IEA (2009) World energy outlook. Technical report, International Energy Agency (IEA)
Immelt J, Govindarajan V, Trimble C (2009) How ge is disrupting itself. Harvard Bus Rev 87(10):56–65
IPCC (2001) Climate change 2001: the scientific basis. contribution of working group i to the intergovernmental panel on climate change. Technical report, Intergovernmental Panel on Climate Change (IPCC)
Lowenstein R (2004) Origins of the crash: the great bubble and its undoing. Penguin Press, UK
MacKay D (2009) Sustainable energy—without the hot air. UIT Cambridge, UK
MEA (2005) Ecosystems and human wellbeing. Technical report, Millennium Ecosystem Assessment (MEA)
Ministry of Finance, India (2010) Union budget and economic survey: budget 2010–2011
Moynihan M, Allwood J (2014) Utilization of structural steel in buildings. Proc Roy Soc A 470
Nature (2010) News briefing: Indian budget. Nature 464:12–13
Nature-Editorial (2021) Editorial: control methane to slow global warming—fast. Nature 596:461
Rao B (2013) How disruptive is frugal? Technol Soc 35:65–73
Rao B (2014) Alleviating poverty in the twenty-first century through frugal innovations. Challenge 57:40–59
Rockström J, Steffen W, Noone K, Persson Å, Chapin FS, Lambin E, Lenton TM, Scheffer M, Folke C, Schellnhuber H, Nykvist B, Wit CAD, Hughes T, van der Leeuw S, Rodhe H, Sörlin S, Snyder PK, Costanza R, Svedin U, Falkenmark M, Karlberg L, Corell RW, Fabry VJ, Hansen J, Walker B, Liverman D, Richardson K, Crutzen P, Foley J (2009) Planetary boundaries: exploring the safe operating space for humanity. Ecol Soc 14(2):32
Steffen W, Persson Å, Deutsch L, Zalasiewicz J, Williams M, Richardson K, Crumley C, Crutzen P, Folke C, Gordon L, Molina M, Ramanathan V, Rockström J, Scheffer M, Schellnhuber H, Svedin U (2011) The anthropocene: from global change to planetary stewardship. AMBIO 40(7):739
UN-Population-Division (1999) The world at six billion. Technical report, United-Nations-Population-Division
World-Bank (2002) Globalization, growth and poverty: building an inclusive world economy. Oxford University Press, New York & the World Bank, USA
World-Bank (2009) The global economic crisis: assessing vulnerability with a poverty lens, policy note. Technical report, World Bank, USA

Chapter 3
Grassroots Frugal Innovations

Grassroots frugal innovations (GFIs) are the oldest frugal innovations to date with origins at the bottom of the pyramid in India and China. In fact, GFIs came into being and were recognized as such in the late 90 s to mid 00 s when India and China were emerging markets of a world revving under globalization. It all started with the quintessential *Jugaad*, a contraption serving the needs of the people at the bottom of the pyramid. Figure 3.1 shows a *Jugaad* in the form of an improvised *cappuccino* maker. Here a pressure cooker has been utilized to retrofit mundane coffee-making into low cost and on the go *cappuccino*. There are several such applications of *Jugaad* which the interested reader can find in these resources (Rao 2013) and others. It should be noted that the term *Jugaad* is used in Northern India for any informal activity that makes a contraption/service successful usually for a brief period of time. As such, there is a negative connotation associated with *Jugaad* in India that has in some cases impeded the growth of frugal innovations. It is the aim of this book to dispel doubts in the minds of our readers and thereby put frugality at the center of sophisticated engineering. Other than the typical *Jugaad*, many other GFIs have arrived over the years and have become famous under other names. Examples include *MittiCool* and *pest-repellence*, earthern refrigerator or fridge and a natural repellent, respectively. This chapter highlights some of these GFIs and also brings out the ingenuity underlying these innovations which are popular among their users. In so doing, it also presents GFIs as a concept that should be utilized along with their advanced brethren in fighting against various maladies that afflict the bottom of the pyramid.

© The Author(s), under exclusive license to Springer Nature Singapore Pte Ltd. 2024
B. C. Rao, *Frugal Engineering*, Design Science and Innovation,
https://doi.org/10.1007/978-981-99-9700-8_3

Fig. 3.1 A *Jugaad* coffee-maker for serving delicious cappuccino on the go

3.1 Ingenuity of Grassroots Innovators

The advent of GFIs and, *frugal innovations* in general, would have been impossible without their creators, the clever and also enterprising innovators among the low-income section of emerging markets. These workers who typically do not possess access to important resources, such as knowledge and finance, are very crucial to the development and also diffusion of GFIs. These workers are talented individuals who possess the inventiveness to innovate out of situations that are naturally scarce on resources. The advent of the *frugal innovation* movement and its continuing progress is a testimony to this group of gifted innovators, who without tapping into formal knowledge bases could achieve their makeshift goals through their dint of both ingenuity and hard work. The creative talents of these individuals could be harnessed to a much larger degree by educating them and/or making timely availability of subsidized knowledge bases.

3.2 Some GFIs

A GFI is typically created to satisfy a need in scarcer settings. In other words, *grassroots frugal innovations* are created against a natural constraint on resources going into their making. Therefore, entrepreneurs at the bottom of the pyramid have to be quite clever and inventive in innovating with lesser materials and/or the absence of a relevant knowledge base. The "knowledge base" here refers to any set of formally laid out principles that are necessary for the working of GFIs. So science, engineering and economics are all knowledge bases relevant to proper GFIs. Such knowledge bases underlying these innovations would significantly aid in creating and also fine-tuning them. Such knowledge bases might be altogether absent or there might be a lack of access due to limited wherewithal including financial. So GFIs are makeshift innovations that have been cleverly assembled or built to carry out their functionality against a perceived need. The *frugal* movement started with these innovations and the significant attention they garnered, due to their astonishing simplicity and good functionality through ingenuity, have catapulted these innovations into advanced settings. Although these innovations do not typically make use of knowledge bases such as science and engineering in a formal manner, their working is rooted in sound principles. Therefore, these cleverly built contraptions are rooted in a rigorous knowledge base even though their clever creators might not be aware of their working. This leads to the interesting observation of the possibility of leveraging *frugal innovations* to a much larger degree through a proper understanding of the underlying knowledge bases by all the stakeholders involved in their creation.

However, setting aside the active involvement of science and engineering for the moment, it would be instructive to describe some of the GFIs in use. This will help understand the concepts underlying GFIs and *frugal innovations* in general. In this regard, this section presents two *grassroots frugal innovations* with more examples to the interested reader being found in the works of Professor Anil Gupta of the *Indian Institute of Management* (IIM A) (Gupta 2013). Moreover, it should be noted that this book's emphasis is on the systematic creation of *advanced frugal innovations* from scratch, the topic of the next chapter and, hence this chapter gives a glimpse of the interesting world of GFIs which undoubtedly are the basis to *frugal innovations* in general.

3.2.1 MittiCool

The *MittiCool* fridge is a good representative of GFI, which is shown in Fig. 3.2. *MittiCool* is completely made from *Mitti*, in the Indian context, or terracotta and as such represents earthenware put to good use by an ingenious creator. The creator, Mansukhbhai Prajapati, is an excellent innovator who tinkers or experiments with his innovations to hone them for an application (Raturi 2010). The *MittiCool* resulted after three years of experimenting with various designs to come up with an earthen refrigerator that does away with modern electrical sources of energy. Instead, it uses the cooling power of evaporation to keep its contents including produce and milk both fresh and cool for a couple of days (Raturi 2010). *Mansukhbhai's* entrepreneurship, inventiveness and successes have invariably led Indian design organizations such as the *National Institute of Design* helping him get formally introduced to systematic design and also research (Raturi 2010). Such formal initiatives give advanced skills which coupled with inherent ingenuity would enable entrepreneurs like *Mansukhbhai* to achieve their goals in a planned and not makeshift manner.

Fig. 3.2 The MittiCool clay refrigerator created by Masukhbhai Prajapati

3.2.2 Improvised Repellant for Pests

This is not a popular example but highlights a key feature of GFIs being makeshift in nature. Accordingly, Professor Anil K. Gupta of the *Indian Institute of Management at Ahmedabad* (IIMA), India (Gupta 2013) discusses the case of farmers in India planting coriander around chickpeas in fields to repel pests. Although the coriander did a splendid job of driving away the pests, the reasons ascribed for the mechanism by the farmers were not scientific. In particular, it was the nectar of coriander plants that attracted the pests according to scientists at the *International Crops Research Institute for Semi-Arid Tropics* in Hyderabad, India and not some pest-repellant housing in the coriander plants. This example clearly highlights the big difference systematic inclusion of knowledge bases, in this case agricultural-science, can make to the effective utilization of *frugal innovations*. The farmers were lucky in this case that the solution worked but for the wrong reasons!

3.3 Drawbacks of GFIs

Grassroots frugal innovations comprise a befitting solution to several problems confronting humanity. Their diffusion is hindered mainly due to their premature failures and/or limited life that result from an undertaking not backed by rigorous application of the relevant knowledge base. Therefore, cleverly putting together a brilliant contraption without due consideration of the underlying science would leave holes in the performance of these products. The makeshift nature of these innovations is clear from the two examples presented in this chapter.

In fact, such makeshift nature of GFIs could jeopardize the diffusion of *frugal innovations* in general. This is because quality of performance is a better metric of market-size in the long run provided customer-surveys and other studies have shown a promising necessity for the given product. In this regard, GFIs may score badly plausibly due to bad performance stemming significantly from the non-involvement of relevant knowledge bases in design, fabrication and their testing. And formal involvement of knowledge bases becomes crucial in sectors like healthcare and defense where functionality is highly dependent on a proper understanding of the science underlying their products.

3.4 Future of GFIs

Grassroots frugal innovations mark the beginning of the entire movement of going *frugal*. GFIs have helped in the advent of *advanced frugal innovations* and subsequently *frugal engineering*. *Grassroots frugal innovations* are significant for the poor and rich alike. These innovations have been and will continue to be important in improving the living standards of society at large. Not only are GFIs crucial to people living at the bottom of the pyramid in both developed and developing economies, their usefulness and/or underlying concepts are also vital to the rich globally. In particular, some of the GFIs could lead to ideas for the engineering of advanced products that could be used in the society at large.

A possible solution to enhancing their diffusion is transforming them into an advanced version by allowing free or subsidized access to the relevent knowledge bases and also workers possessing advanced skills in those areas. Furthermore, the creators of *advanced frugal innovations*, the topic of the next chapter, need to frequently tap into the domains of GFIs to possibly borrow ideas and even a GFI that could be engineered for sophisticated applications.

References

Gupta A (2013) Tapping the entrepreneurial potential of grassroots innovations. Stanford Soc Innov Rev 11(3):18–20
Rao B (2013) How disruptive is frugal? Technol Sol 35:65–73
Raturi P (2010) 'Potter power: son of the soil

Chapter 4
Advanced Frugal Innovations

4.1 Introduction

The phenomenon of *grassroots frugal innovations* sheds light on the potential of *frugal innovations*, in general, as cost-effective and environment-friendly products with good functionality. However, the absence of a formal design and development is a major hindrance for GFIs since makeshift products typically do not work at their best for long. In contrast, planned and meticulous design and fabrication of *frugal innovations* would significantly aid in realizing and hence harnessing their full potential. Such methodologies for creating *frugal innovations* from scratch require a proper understanding of the physics underlying their working.

Recent years have seen the advent and widespread diffusion of a new type of *frugal innovation* called *advanced frugal innovation* (AFI). AFIs are sophisticated products created through the systematic application of a knowledge base such as *science and technology* (S&T) (Rao 2017a, b). Besides S&T, other non-science-based knowledge bases could also be tapped into depending on the area of application of the AFI. However, this book will refer from here onwards to applications in science and engineering and the term *science & technology* (S&T) will be invoked in alluding to these applications. However, the reader is entreated to think about AFIs in other fields, and in the process, think about a "universal" knowledge base encompassing ideally all the domains of knowledge available to denizens of earth. Such a *universal* knowledge base should include sciences, engineering, technology, economics, finance, management science, history, geography and all other spheres of knowledge available on the face of the earth. In other words, all-inclusivity of the knowledge base would account for all the modern spheres of knowledge and also the interactions between various fields to consider interdisciplinary areas. Armed with such a knowledge base, innovators would "frugalize" vast swathes of human endeavors. So against this backdrop, let's proceed to a formal definition of an AFI and also study its various aspects for effective utilization.

© The Author(s), under exclusive license to Springer Nature Singapore Pte Ltd. 2024

B. C. Rao, *Frugal Engineering*, Design Science and Innovation,
https://doi.org/10.1007/978-981-99-9700-8_4

4.2 Definition

The defining features of an *advanced frugal innovation* are the following:

a. Their cost is low due to their frugal nature.
b. They leverage advances in S&T from the incipient stage. *Science & technology*, as used in this effort, encompasses the use of principles and/or advances in sciences, engineering, technology, mathematics and any other knowledge-based area relevant to a given AFI. Unlike their grassroots brethren that depend on serendipity and/or tinkering for creation (Gupta 2013), AFIs need significant involvement of S&T from the incipient stage to account for the sophistication in their designs.
c. They are undertaken by workers with advanced degrees—see Sect. 6.1—who can leverage advances in S&T for achieving an AFI.

4.3 Quintessential Example of AFI

A quintessential example of an AFI is the *Gravitylight*^{TM} (Rao 2017a), which uses earth's gravity to power a light source, thus making it a widely applicable low-cost lightning solution in remote and far-flung regions of the world housing disadvantaged sections of society. In fact, *Gravitylight*^{TM} could also find applications in developed or richer countries where it is a sustainable low-cost alternative for many applications. *Gravitylight*^{TM} clearly displays a good theoretical and working knowledge of gravity on part of the creators to make this product a commercial success. *Gravitylight*^{TM} is cost-effective, compact and sophisticated despite its simplicity which attests to the formal involvement of creators well versed in S&T for its successful realization. Figure 4.1 shows the sophisticated innards of a *Gravitylight*^{TM}.

Figure 4.2 shows another AFI that powers water filtration through the pedaling action of a human. This product, which is a work in progress in the author's group at IIT Madras, can be utilized in both poor and rich settings for having quick access to potable water. This device has been designed from scratch using off-the-shelf filters and also knowledge of pumping action. Here again the innovation is planned from scratch against the constraint of low cost needing formal design and engineering principles for its fruition. In fact, what was planned for an undergraduate project thesis has turned out to be a complex problem needing a good understanding of science and engineering to make the original low-cost idea possible.

An important caveat here is that an AFI does not sacrifice safety and other features for achieving low cost. Such misconceptions will actually result in an innovation that is NOT an AFI! The challenge is to achieve safety in the quality functionality at low cost with minimal resources. In other words, we have to strive to making products (as AFIs) that are as good or better vis-à-vis the current crop of bulky products prized for their safety and performance.

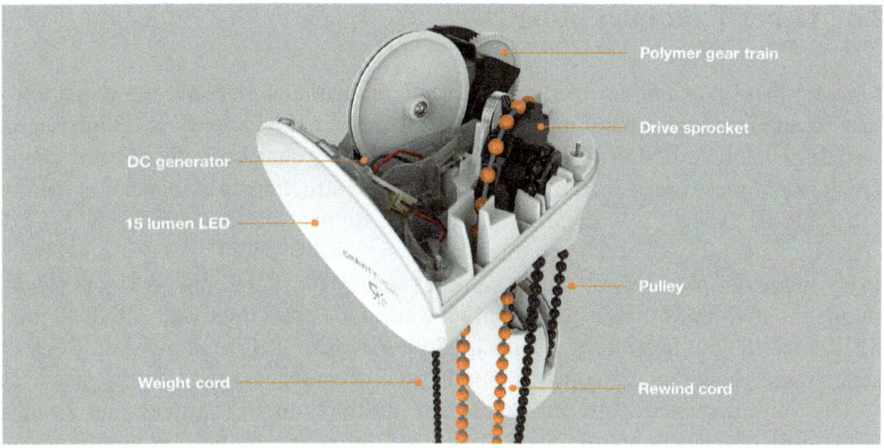

Fig. 4.1 Inner workings of an quintessential AFI, *Gravitylight*^{TM}. Image is taken from https://deciwatt.global/gravitylight

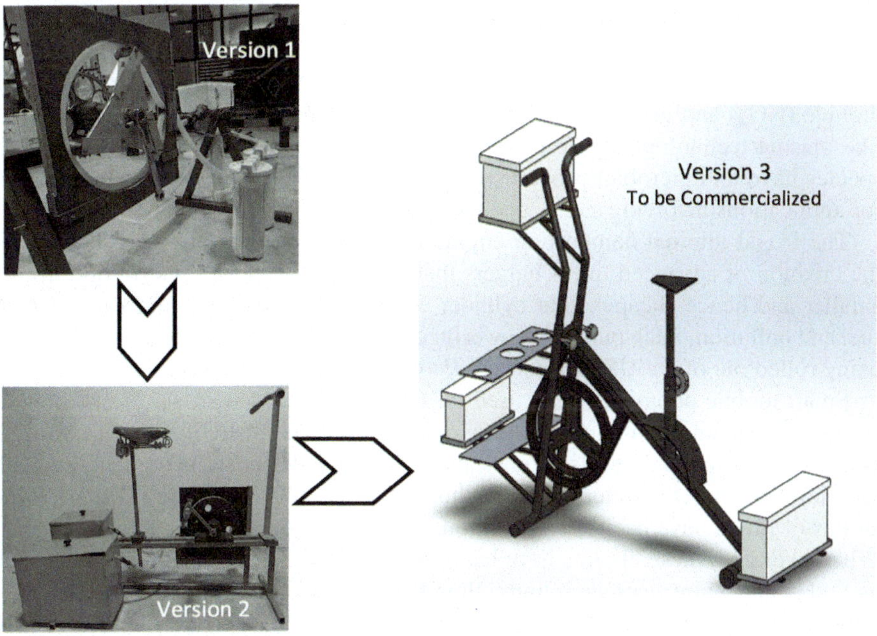

Fig. 4.2 An *advanced frugal innovation* designed for water filtration (Rao and Bhinge 2013). Image for Version 3 is taken from Rao (2017b)

4.4 Other Examples of AFIs

Table 4.1 lists 25 examples of advanced frugal innovations that have appeared in the past decade in sectors spanning from cars to healthcare, weaponry and high energy particle physics. Although all of them share the primary characteristic of being cheap, a majority of these are also compact with a no-frills structure as evidenced in column 5 of Table 4.1. The AFIs with no entries under "Compact Design" are deemed so due to the implausibility of creating a compact version with Sandia's robotic hand being an obvious example. Overall, these examples clearly attest to doing good science and engineering with less thereby eventually lowering the costs involved. A look at more AFIs since the publication of entries in Table 4.1 (Rao 2017a) is given in Sect. 4.4 under Table 4.2 with detailed explanations for each entry.

The rainfall sensor (Tollefson 2013) and plasma collider (Downer and Zgadzaj 2014) utilize newer low-cost methods backed by research such as lightning data for rainfall estimation (Tollefson 2013) and a 30 cm plasma afterburner for boosting electrons coming from a conventional accelerator (Downer and Zgadzaj 2014), respectively. Traditionally, rainfall prediction uses expensive and bulky Doppler radar while the particle colliders are costly behemoths smashing particles that are accelerated over distances spanning several kilometers.

Continuing with plasma, Pavlovich (2014) has reported the production of nitrogen dioxide (NO_2) and nitric oxide (NO) species in the air through inexpensive spark-like plasma technology that runs on low-cost power supplies. The NO_2 and NO species have antimicrobial properties that make this frugal plasma technology apt for applications involving sterilization and antisepsis, to name a few.

The frugal internal combustion engine is another instance of research used for the creation of advanced turbochargers that deliver more power and torque from a smaller and hence cheaper, four cylinder engine, while also economizing on both fuel and pollution. Such puny but powerful engines are used in the luxury car models being rolled out of the US and Europe (Economist 2015).

Smart mobile phony has been used for the low cost and successful screening of communicable and non-communicable diseases such as tuberculosis and intestinal worms. These applications use probes and/or inexpensive lenses in conjunction with mobile phony for the detection of pathogens (Joyce 2012a; Nature 2013). Some strains of Ebola have been detected through dehydrated gene circuits laid on paper. When hydrated, these circuits have been successful in detecting Ebola strains such as Sudan and Zaire (Research highlights: Paper-based gene tools 2014). Recently, a low-cost palm-sized DNA sequencer, and its variats, have been developed whose coupling with mobile phony in the foreseeable future would allow fast identification of unknown fauna (Hayden 2015; Quick et al. 2016). All of these mobile phony-based AFIs are compact when compared to the bigger setups housed in clinics with the supporting paraphernalia.

Cloud technology has been utilized by a group of entrepreneurs in communicating data with a low-cost sensor fabricated by 3D printing. The sensor called Miriam analyses blood samples looking for suitable microRNA markers for cancer that are

Table 4.1 Examples of *Advanced Frugal Innovations*

No	Advanced Frugal Innovation (AFI)	Manufacturer (Country)	Frugal Features	Compact Design	Price
1	Light (http://gravitylight.org/)	GravityLight (GravityLightTM, UK)	Using the descent of a weight under gravity to generate 25 minutes of lighting without electricity	Yes	£3
2	Rural Electrification (Jeff 2014)	(US and India)	Uses cheap microgrids and smart meters to deliver eco-friendly electricity to rural India at a lower cost	–	–
3	Car (Economist 2010)	Tata Motors (Tata Nano®, India)	Basic "no frills" car that was built to minimize unnecessary features	Yes	$2200
4	Frugal Four Cylinder Engine (Economist 2015)	(USA and Europe)	Use of advanced turbochargers for amplifying the power and torque of four cylinder engines while providing economy in both fuel and pollution	Yes	–
5	Electrocardiogram (Immelt et al. 2009; Economist 2010)	General Electric (Mac 400® &Mac 800®, India)	ECG reduced to a handheld device with a built in small printer	Yes	$800
6	Ultrasound (Immelt et al. 2009)	General Electric (China)	Portable machine consisting of a laptop computer with sophisticated software and associated probe	Yes	$15,000–$100,000
7	Speech synthesizer (Technology-Review 2011)	Invention Labs (Avaz®, India)	Compact unit handling multiple languages made cheap by working directly with Asian hardware manufacturers	Yes	$800
8	TB Probe (Joyce 2012b)	Stanford University (US)	Probe fluorescent molecule when cleaved by a TB protein releases faint blue color picked up by a smartphone camera and later transmitted to clinicians for frugal diagnosis	Yes	–
9	Identification of Intestinal Worms (Nature 2013)	Toronto General Hospital (Canada)	An inexpensive lens attached to an iPhone allows low-cost optical micrsocopy of stool samples	Yes	–
10	Identification of Ebola (Research highlights: Paper-based gene tools 2014)	Boston University (USA)	Gene circuits on paper for identifying pathogens	Yes	–

(continued)

Table 4.1 (continued)

No	Advanced Frugal Innovation (AFI)	Manufacturer (Country)	Frugal Features	Compact Design	Price
11	Frugal Cancer Detection (This device could detect dozens of cancers with a single blood test 2014)	Miroculus (Miriam, US)	Monitors apt microRNA in blood sample through a compact 3D printed device that communicates with the Cloud	Yes	–
12	Portable DNA Sequencer (Hayden 2015)	Oxford Nanopore Technologies (MinION, UK)	Palm sized gene sequencer that plugs into a laptop USB for displaying data online instead of at end of run	Yes	–
13	Robotic Hand (Greenemeier 2012)	Sandia Labs (US)	Improved dexterity mimicking real-time gestures through an electronic system comprised of cell-phone-parts	–	$10,000
14	Swarms of Simpler Robots (Nature 2014)	Massachusetts Institute of Technology (Kilobots, US)	Swarms of low-cost simple robots perform complex tasks in unison	Yes	–
15	Robotic Submarine (Davis 2014)	Falcon Robotics (Stinky, US)	Underwater robotic submarine fabricated under a shoestring budget wins NASA sponsored Marine Advanced Technology Education Robotics Competition	Yes	$1,000
16	Pollution sensors (Austen 2015)	Do-It-Yourself (DIY) Kit (DustDuino, Air Quality Egg and Smart Citizen Kit, US)	Cheap sensors built through the DIY movement that provides instructions on the internet	Yes	$185 (A.Q. Egg) $50 (DustDuino)
17	Lightning sensor (Tollefson 2013)	Earth Networks (USA)	Lightning data tracked by sensors on top of mobile towers are converted to rainfall estimates by intricate software	–	$1,000,000
18	Missile (Economist 2012)	Naval Air Weapons Station (Spike, US)	Low-cost small missile made from off-the-shelf components. This has an ingenious optical guidance system	Yes	$5,000

(continued)

Table 4.1 (continued)

No	Advanced Frugal Innovation (AFI)	Manufacturer (Country)	Frugal Features	Compact Design	Price
19	African Radio Telescope (Nordling 2012)	(Square Kilometer Array, South Africa)	Satellite dishes used for telecommunications are retrofitted to create low-cost radio telescopes	–	R15,000,000–20,000,000
20	African Telescope for Spectroscopy (Nordling 2013)	(South African Large Telescope (SALT), South Africa)	Design induced by a shoestring budget to build a telescope with minimal features for accurate spectroscopy	–	\$30,000,000
21	Indian Martian Spacecraft (Economist 2014; Kumar 2014)	Indian Space Research Organization (Mangalyaan, India)	The spacecraft achieves its frugality through launch by a low-power rocket and also by carrying fewer sensors and instruments	Yes	\$74,000,000
22	Smaller satellites (Jones 2014)	(CubeSats, US)	Small and compact satellites comprised of cheap parts that are also launched free of cost	Yes	<\$100,000
23	Reuse of Rockets for Space Flight (Scientific American 2015)	SpaceX (Falcon 9, US)	Successful in launch and safe reuse thus reducing costs of the first stage Falcon 9 booster rocket	–	–
24	Plasma Particle Accelerator (Downer and Zgadzaj 2014)	SLAC National Accelerator Laboratory (FACET, US)	Use of plasma afterburner about 30 cm long has resulted in a low-cost technology capable of boosting electrons from a conventional accelerator 500 times	Yes	–
25	Frugal Plasma Technology (Pavlovich 2014)	University of California Berkeley (Frugal Plasma, US)	Use of inexpensive spark-like plasma technology to create antimicrobial species in air for applications involving sterilization	–	–

Table 4.2 More examples of *Advanced Frugal Innovations*

No	Advanced Frugal Innovation (AFI)	Manufacturer (Country)	Frugal features	Compact design	Price
1	Arduino and Raspberry Pi (Cressey 2017)	Arduino (https://www.arduino.cc/, US) Raspberry Pi (https://www.raspberrypi.org/, UK)	Arduino is a microcontroller and Rasberry Pi is a computer Both are low-cost alternatives for DIY and other projects	Yes	Both are affordable <$100
2	Solo the quadruped robot (Grimminger et al. 2020)	Solo (Max Planck Institute for Intelligent Systems, Germany and (New York University, US)	Solo is a qaudruped robot The low-cost results from the off-shelf items, 3D printed parts and intelligent research	Yes	Cheaper than commercial versions.
3	Drugs for Neglected Diseases initiative (DNDi) (Maxmen 2016)	DNDi (Kick started by Médecins Sans Frontières, Switzerland)	DNDi is a non-profit initiative for making new drugs at a fraction of the outlays of Big Pharma	–	$110–$170 million
4	Impulse Drum Charger (IDC) (Cruccolini et al. 2018; BikeBandit 2018)	IDC (AlterEgo Hardware Italy)	A low cost and ingenious simple charger that after bolting on a small motorbike boosts its power	Yes	Tiny fraction of a turbocharger
5	Smartphone and microfluidic chip (Sun et al. 2020)	Point-of-Care System (University of Illinois, Urbana-Champaign, USA)	Uses smartphone camera to detect presence of animal-based infectious pathogens through fluorescence from a special dye mixed in nasal swab samples	Yes	Tiny fraction of a lab-based standard clunky equipment
6	Smartphone-based TRI analyzer (Long et al. 2017)	Point-of-Care System (University of Illinois, Urbana-Champaign, USA)	A portable spectrum analyzer that can handle multiple biosensing modalities and hence carry out a wide range of body fluid-based tests over a smartphone	Yes	$550
7	Rotary Wankel Engine (Weiss 2016) (Vorraro et al. 2019)	650 Series (Advanced Innovative Engineering (AIE), UK)	Rotary type *internal combustion* engine that is compact and low cost vis-à-vis piston type	Yes	Lower than piston type engines

(continued)

Table 4.2 (continued)

No	Advanced Frugal Innovation (AFI)	Manufacturer (Country)	Frugal features	Compact design	Price
8	*Advanced Wakefield Experiment* (AWAKE) (Gibney 2015; AWAKE 2016)	AWAKE (CERN, Switzerland)	Uses plasma wave through pulses of protons for accelerating charged particles over much shorter distances	Yes	Lower than bulky accelerators spread over several Kilometers
9	Chandrayaan-2 (Padma 2017)	Chandrayaan-2 (Indian Space Research Organisation (ISRO), India)	An ingenious Moon mission involving an orbiter, lander and rover achieved under a low budget	Yes	$93 million Low by space mission standards!
10	Cheap MRI (Nature 2015)	Cheap MRI (Harvard Medical School, US)	Uses small magnets to make tissue emit weaker radio signal imaged by a relevant sensor	Yes	One-twentieth cost of conventional MRI
11	Colloidal quantum dot spectrometer (Bao and Bawendi 2015)	CQD Spectrometer (Tsinghua University, Massachusetts Institute of Technology, California Institute of Technology, China and US)	Spectrometry with *colloidal quantum dots* is smart-phone-based and hence portable and cost-effective	Yes	Cheaper than traditional spectrometers
12	Chipsats (Jones 2016; Space 2019)	KickSat-2, (Cornell University, Ithaca, New York)	Satellites in the form of cracker-sized circuit boards that perform tasks effectively	Yes	About $100
13	POC Cancer Detection (Prasad et al. 2019)	FluoroZen, (Louisiana State University, USA.)	A smartphone based tool for detecting cancer gene BRCA-1 through fluorescence of samples on slides based on microfluidics	Yes	–

(continued)

Table 4.2 (continued)

No	Advanced Frugal Innovation (AFI)	Manufacturer (Country)	Frugal features	Compact design	Price
14	Digital Powertrain for vehicles (Stone et al. 2017; GreenCarCongress 2019)	*Intelligent Valve Technology* (IVT), (Camcon Automotive, UK)	IVT does away with camshaft of traditional IC engines by offering precise digital control of valves through a set of electric actuators.	Yes	Will be cheap in futuristic versions
15	Cheap Seismometers (arsTECHICA 2020)	Android OS MyShake (Google and University of California at Berkeley, USA)	Use of accelerometers in multitudes of smartphones along with Android OS for low-cost seismometery	Yes	–
16	*Open science hardware* (Gibney 2016)	Open science hardware (Europe and USA)	A *open movement* for science hardware. Eg: DIY syringe pump with blueprints for fabrication and setup at *Nature's* website	Yes	Just $100 for single syringe pump
17	*Gas turbine engine cores* (Norris 2020)	Rolls Royce LibertyWorks (USA)	Small gas turbine jet engine cores have been developed for attritable flights with emphasis on drones and possibly missiles. Low cost was a significant consideration during development	Yes	Low cost
18	*3D Fuge* Frugal Centrifuge (Byagathvalli et al. 2019)	Georgia Tech University (USA)	A 3D printed hand-operated centrifuge based on a child's whirligig. Very useful for remote and low-cost operations	Yes	Quite lower than standard benchtop type

(continued)

Table 4.2 (continued)

No	Advanced Frugal Innovation (AFI)	Manufacturer (Country)	Frugal features	Compact design	Price
19	*Gravitricity* Frugal heavy power source (Irving 2020)	Gravitricity (https://gravitricity.com/) (UK)	Use of gravity enabled potential energy of heavy weights to generate heavy-duty power	Yes	Low cost alternative to even renewable types
20	*Flexible Paper Electronics* (Medeiros et al. 2020)	School of Industrial Engineering Purdue University (US)	Flexible electronics are printed on paper or cardboard to create low-cost human-machine interfaces like keyboards Also, no power source is required!	Yes	As low as $0.25 per device
21	*Bio-MOD* Cheap medication on the go (Arnold 2019)	Chemical, Biochemical and Environmental Engineering University of Maryland Baltimore County (US)	Making sophisticated drugs on the go through miniaturized drug production system in a briefcase!	Yes	Low cost
22	*Frugal fertilizer* Agriculture without fertilizers (Gilbert 2016)	Pennsylvania State University University Park (US)	Use of natural plant features like suitable roots to improve yield in nutrient starved soils	–	Low cost
23	*LoCHAid* Really low-cost hearing aid! (Lesté-Lasserre 2020)	Georgia Institute of Technology (US)	A really low-cost hearing aid developed from off-the-shelf components	Yes	Retail: $1 Fabricating in-house from blue-print: $15–$20
24	*Spirometer* Low-cost lung diagnostic tool (Popular Mechanics 2015)	Maya Varma (US)	A really low-cost *spirometer* that detects asthama, lung disease and other ailments. Built through 3D printing and supporting mobile phone app.	Yes	Under $35

(continued)

Table 4.2 (continued)

No	Advanced Frugal Innovation (AFI)	Manufacturer (Country)	Frugal features	Compact design	Price
25	*Gentlewasher* Frugal washing machine (Shah 2018)	Coen Vermeer & Kathak Mehta (Centre for Environmental Planning and Technology (CEPT), Ahmedabad, India and Erasmus University Rotterdam, the Netherlands)	A compact hand cranked washing machine that does a great wash in 4–5 mins	Yes	Low cost in future
26	*African Radio Astronomy* From used communications dishes (Wild 2017)	*African VLBI Network* (AVN) (African initiative)	Retrofitting discarded telecommunications dishes for radio astronomy	Yes	Low cost (similar to entry 19 of Table 4.1)
27	*VIRRION* POC viral detection from nanotechnology (Yeh et al. 2020)	*The Pennsylvania State University* and *New York University* (USA)	A tiny device using *carbon nano tubes* to trap viral particles with subsequent analyses by Raman spectroscopy to find viral signatures within minutes	Yes	Low cost
28	*Low-cost Intraocular lens* (IOL) Low-cost lenses for eye care (Popular Mechanics 2016; Aurolab 2024)	*Aravind Eye Care System* and *Aurolab of Aravind* (Tamil Nadu, India)	Researching into original eye products. Started from ingenious reverse engineering	Yes	Low cost
29	IKEA (2022); Lee (2019)	IKEA	Single piece of furniture is multifunctional. Ideal for cramped spaces. Possibility of making this robotic	Yes	Affordable for urban micro-space dwellers
30	ClimeCell (Tollefson 2017)	*ClimeCell* and European and Israeli companies (Europe, Israel and US)	Approach uses microwave signal strength to forecast street level precipitation	Yes	Low cost
31	*OpenRAN* Virtual 5G Network (Economist 2020)	*Rakuten* (Japan)	Majorly software based 5G network. Cost savings coming from lesser hardware	–	Inexpensive vis-à-vis traditional 5G infrastructure

(continued)

Table 4.2 (continued)

No	Advanced Frugal Innovation (AFI)	Manufacturer (Country)	Frugal features	Compact design	Price
32	*Redox flow batteries* (Janoschka et al. 2015)	*Friedrich Schiller University, Center for Energy and Environmental Chemistry, JenaBatteries GmbH* (Germany)	A Redox flow battery technology that uses inexpensive dialysis membranes along with climate-friendly electrolyte. Possesses good energy density	–	Inexpensive vis-à-vis traditional counterparts
33	*Bio-inspired Image Sensor* (Garcia et al. 2018)	*Washington University, Southern Illinois University at Edwardsville, University of Arizona, Veterans Affairs St. Louis Health Care System,* and *University of Illinois at Urbana-Champaign* (USA)	Bio-inspired from Morpho butterfly, this image sensor aids clinical imaging during surgery of cancerous tissues Is effective even in a poorly illuminated surgery-room	–	Quite low vis-à-vis traditional equipment
34	DIY Confocal Microscope (Ravindran 2020)	Federal University of Minas Gerais in Belo Horizonte (Brazil)	Low-cost sophisticated confocal microscopy through plexiglass stage and infrared lamp both of which bought from local stores	Yes	Very low!
35	DIY Optical Microscope (OpenFlexure) (Ravindran 2020)	University of Bath (UK)	Super-resolution microscope made by 3D printing OpenFlexure design available for public use	Yes	$262
36	Solar based Autoclave (Zhao et al. 2020)	*Massachusetts Institute of Technology* (US)	Ingenious sterilization of surgical instruments through saturated steam generated through solar heating of fins whose losses are minimized through an aerogel	–	Just $38

(continued)

Table 4.2 (continued)

No	Advanced Frugal Innovation (AFI)	Manufacturer (Country)	Frugal features	Compact design	Price
37	Machine Learning for Cancer Detection (Esteva et al. 2017)	*Stanford University* (US)	Use of *machine learning* for training and subsequent use of a computer system for initial screening of skin cancers. Low-cost prediction at par or better than qualified doctors	–	Low cost
38	Artificial Eye for Scotopic Vision (Liu et al. 2016)	*University of Wisconsin-Madison* (Madison, US)	Use of *biologically inspired photosensitivity enhancer* (BPE) comprising multitudes of microphotocollectors (μ-PCs) for enhancing photosensitivity of low-imaging applications. Low cost and consumes zero power!	–	Low cost
39	*Gravimeter* (Middlemiss et al. 2016)	*Scottish Universities Physics Alliance (SUPA), University of Glasgow* (UK)	MEMS-based *gravimeter* comprises two anti-flexures and also a cantilever-flexure etched out of silicon to constrain stable motion of proof-mass in one direction at low frequencies	–	Very low cost!
40	*OneWeb and Airbus* https://onewebsatellites.com/	*Oneweb and Airbus* (Florida, US)	Mass production of modular satellites for sophisticated applications	–	Low cost
41	*Tribot* (Zhakypov et al. 2019)	*Swiss Federal Institute of Technology Lausanne, Osaka University* (Switzerland and Japan)	Bio-inspired multi-locomotion robot for diverse collective tasks. Built out of PCB technology and mass production to keep costs low	–	Low cost $60 a bot
42	CMOS-enabled energy harvester for low-cost IoT (Tarancón 2019)	*University of Texas at Dallas, Texas Instruments Inc* (US)	CMOS-enabled nanostructured silicon harvesters are being designed for harvesting energy from heat dumped into the surroundings by a device as heat. This is a long-term low-cost solution for replacing batteries. This will make IoT an affordable technology	–	Low cost

(continued)

Table 4.2 (continued)

No	Advanced Frugal Innovation (AFI)	Manufacturer (Country)	Frugal features	Compact design	Price
43	Paper-based sensor to diagnose Zika (Pardee et al. 2016)	*Massachusetts Institute of Technology, Harvard University, Arizone State University, Boston University, Cornell University, University of Wiscosin-Madison, University of Toronto* (Canada and US)	Modified paper-based sensor for Ebola to aid detection of Zika virus. Uses RNA-sensors programmed and embedded in strips of paper. These strips get bonded to viral sequences leading to a reaction that changes paper-color. Significant in fighting pandemics of Zika!!	–	Low cost
44	Healing Patella Dislocation (Dey et al. 2016)	*University of Cape Town* (South Africa)	A low-cost drill-guidance device fabricated through 3D printing. By guiding drills to exact location for MPFL surgery this device avoids the dangers and high-cost of X-ray imaging used currently for locating region for surgery.	–	Low cost. Just ~$20 a pop
45	*Edokko Mark I* Deep Ocean Observatory (Nature 2021)	*Research and development partnership for next generation technology of marine resources survey* (J-MARES) (Japan)	A low-cost deep ocean observatory. Affordability stems from fabrication of glass spheres with very precise sphericity and associated finish and texture	–	Low cost
46	Disinfecting Water with Sunlight (Liu et al. 2016)	*Stanford University* (USA)	Uses visible spectrum of sunlight for the photocatalytic disinfection of water. Vertically layered *molybdenum disulphide* (MoS2 or FLV-MoS2) films together with Cu and Au depositions for harvesting visible light that is subsequently used to efficiently disinfect water	–	Low cost. And fast!
47	*OneBreath* Mechanical Ventilator (Stanford 2024)	*Stanford University* (USA)	A ventilator made out of low cost components. Its quality performance is designed for resource-poor regions and emergencies in other settings	–	Low cost

(continued)

Table 4.2 (continued)

No	Advanced Frugal Innovation (AFI)	Manufacturer (Country)	Frugal features	Compact design	Price
48	Bamboo Housing *CUBO* (BBC-News 2018)	CUBO Communal Housing (Philippines)	Use of engineered bamboo for building sophisticated housing of different types	–	Low cost
49	*optical coherence tomography* (OCT) Scanner (Song et al. 2019)	Duke University (USA)	Built out of low-cost parts. Ingenuity lies in a 3D printed plastic-spectrometer that makes light travel in self-correcting circular paths as opposed to linear ones of traditional OCTs	–	Low cost. ~$15,000
50	*Microscopy with LEGO®* (Vos et al. 2021)	Münster University and Georg August University Göttingen (Germany)	Low-cost optical microscopy through a LEGO®-based microscope also using off-the-shelf low-cost but powerful lenses	–	Low cost. ~ €80
51	*Additive Manufacturing* (AM) (Scientific American 2016)	*OpenBCI and Department of Energy (DOE)* (USA)	OpenBCI developed low-cost brain-computer interface with 3D printed headset or skullcap. DOE 3D printed, at low cost, 30 m long wind turbine blades with intricate patterns for improving efficiency	–	Inexpensive vis-à-vis traditional counterparts
52	*Advanced Frugal Innovator* (https://www.macfound.org/*fellows/class-of-2016/rebecca-richards-kortum*)	*Rice University* (USA)	Professor Rebecca Richards-Kortum, who won the prestigious genius-grant by the MacArthur Foundation, specializes in low-cost diagnostic techniques for resource-poor regions with an example of her AFI being a high-resolution microendoscope for very-affordable treatment of cancer	–	Low cost!
53	*"Mini" drug factory!* (Nature 2019)	*CONTINUUS Pharmaceuticals* (Massachusetts, USA)	Mini factory which is apparently a smaller version of conventional drugs factory. This factory makes a generic drug cheaply and also saves significantly on energy while doing so. Quality is similar to those made conventionally	–	Low cost!

(continued)

Table 4.2 (continued)

No	Advanced Frugal Innovation (AFI)	Manufacturer (Country)	Frugal features	Compact design	Price
54	On-Chip Accelerator (Sapra et al. 2020)	*Stanford University* (USA)	A miniaturized accelerator for producing high-energy electron beams in special channels on a chip. This will pave the way for new applications in biomedicine	–	Significantly inexpensive vis-à-vis the kilometers -long traditonal type
55	Foldscope: Paper Microscope (Cybulski et al. 2014)	*Stanford University* (USA)	An ingenious paper microscope that uses *Origami* to develop *Foldscope*. This very compact microscope uses just thumbing, etc., for operation. Useful in science and education	–	Just $1.0!
56	iPhone Measures Erosion (Leeming 2022)	*University of Copenhagen* (Denmark)	Researcher uses an iPhone with builtin LIDAR for measuring coastline erosion	–	Low cost
57	Cheap and Green Aluminum (Nature-Editorial 2018)	*Alcoa and Rio-Tinto* (USA and Australia and UK)	Combination of green-anodes and green-power to eventually make aluminum production cheap and green	–	Low cost
58	Embodied Logic (Jiang et al. 2019)	*University of Pennsylvania, PA* (USA)	Self actuating structures needing no artificial powering and sensors, etc., for their functioning. Examples include artificial flytrap that closes on its own when a heavy load is introduced	–	Low cost
59	Minimally Invasive Surgery (Awtar et al. 2010)	*University of Michigan, MI* (USA)	An all-mechanical surgical tool for minimally invasive surgeries that has a mechanism for following the surgeon's wrist movements	–	Just $500!
60	Low-Cost Aircraft Wing (Airforce-Technology 2021)	*Continuous Composites* (USA)	A patented *additive manufacturing* technique prints low-cost spars for aircraft wings	–	Low vis-à-vis traditional expensive manufacturing

(continued)

Table 4.2 (continued)

No	Advanced Frugal Innovation (AFI)	Manufacturer (Country)	Frugal features	Compact design	Price
61	Low-Cost Geothermal Energy (Miller 2021)	*West Virginia University* (USA)	A *3D Printed sparger head* for raising geothermal fluids using high-pressure bubbles for cost-effective energy generation	–	Low vis-à-vis traditional expensive pump systems
62	Low-Cost Laser Weaponry (Extance 2015)	*Boeing* (USA)	*Boeing's High Energy Laser Mobile Demonstrator* uses cheaper but compact fiber lasers for effective shots at military targets	–	One shot at $10!
63	TOLIMAN Telescope (Tuthill et al. 2018; UnivofSydney 2021)	*University of Sydney* (Australia)	A low-cost telescope that will detect exoplanets near to our solar system affordably. The agile and precise detections are made possible with a new lens leading to complex floral image-patterns containing the desired signal	–	Low vis-à-vis expensive ones
64	*Sunflower* Solar (Nature 2022)	*Tulane University* (USA)	Solar cells arranged at center of *Sunflower* pattern for efficiently capturing sunlight. Waste heat converted to additional electricity. Gives about 65% efficiency	–	Low vis-à-vis hybrid ones
65	Iodine Propulsion (Rafalskyi et al. 2021)	*ThrustMe* (France)	Low-cost propulsion for space applications based on iodine. Performance is better and cheaper than those using Xenon. Satellites based on this propulsion are also simpler and cheaper	–	Low vis-à-vis Xenon systems
66	Compact Gamma-ray Monitor (Gibney 2021)	*RIKEN* (Japan)	Cheap, compact and portable sophisticated table-top detectors to forewarn dangerous radiation of Gamma-ray bursts coming from thunderstorms on Earth	–	Cheap vis-à-vis conventional systems
67	Biodegradable Robotic Actuator (Heiden et al. 2022)	*Johannes Kepler University* (Austria)	Inexpensive and biodegradable soft robotic actuator that precisely tackles obstacles in its path	–	Low cost

(continued)

Table 4.2 (continued)

No	Advanced Frugal Innovation (AFI)	Manufacturer (Country)	Frugal features	Compact design	Price
68	*Point-of-Care Devices* (Miesler et al. 2020)	*University of Stuttgart* (Germany)	Professor Alexander Brem has clearly brought the inherent AFI features in general *point-of-care devices*	–	Low cost
69	*Micro Electric Vehicle* (techcrunch 2021)	*Eli Electric Vehicles* (USA)	A *micro electric vehicle* called ZERO allows personal mobility affordably with cutting-edge technology	–	Low cost
70	Wooden Satellite (Economist 2021)	*Sumitomo Forestry* (Japan)	Satellites made of a new composite, a form of Wood, will be cheap due to Wood and possess a simple design again due to Wood!	–	Low cost
71	Window Shading (Kay et al. 2022)	*University of Toronto* (Canada)	Window shading mechanism borrowed from ocean-based *Krill* uses pigment-based solution pumped for shading from sunlight. Saves 30% energy	–	Low cost
72	Tesla Heat Pump (Arenaev 2023)	*Tesla* (US)	A compact and simple heat pump for heating car-cabin designed as a single unit. This pump does away with 15 plus components and handles the worst of winters. Saves a lot on energy!	–	Low cost
73	Miniaturized Microscope (Hai et al. 2023)	*Northwestern University* (US)	Mini microscope 8 mm in width built out of 3D printing in just 50 mins at US$4!	–	About US$4!
74	Cuban Biomedical Centres (Reardon 2016)	*Center of Molecular Immunology* *Center for Genetic Engineering and Biotechnology* *Center for Immunoassays* neurosciences center (CNEURO) (Cuba)	Developed world class vaccines, and treatments for cancer and other disorders at low cost. Also, a $2 hearing aid for children	–	Inexpensive vis-à-vis rich world counterparts
75	Pollution Sensors (Lewis and Edwards 2016; deSouza et al. 2017)	(United Nations Environment Program (UNEP) Executed in Nairobi, Kenya)	Low-cost pollution sensors deployed to kindle awareness for pollution control and also supplement traditional sensors	Yes	About $1500

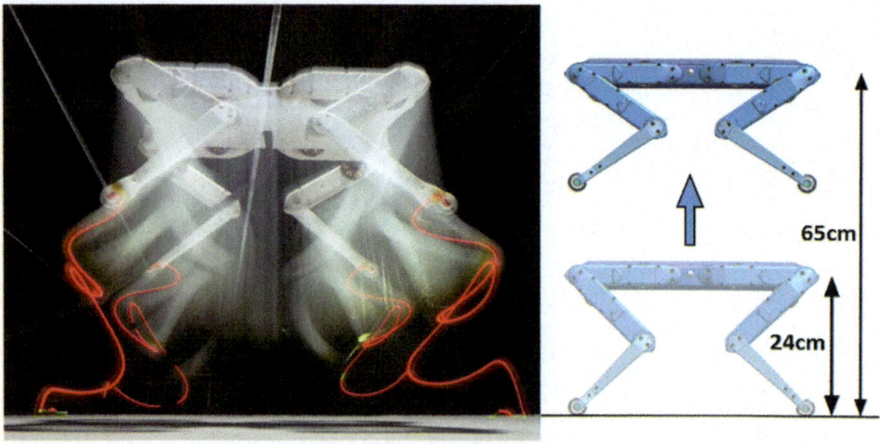

Fig. 4.3 *Solo* a low-cost quadruped. Image is taken from Grimminger et al. (2020) and photo of the jumping robot was taken by W. Scheible

subsequently communicated to the biochemistry-based algorithms housed in the Cloud that aid in the fast and cheap but smart detection of different types of cancers (This device could detect dozens of cancers with a single blood test 2014). The use of a cheap 3D printed sensor together with the Cloud technology makes this application frugal.

The proliferation of *do it yourself* (DIY) kits has unleashed cheaper sensors for tackling pollution that is being deployed by multitudes of mostly amateur scientists. The big-data streaming from myriads of these sensors has facilitated monitoring of pollution particulate that are 2.5 μm (Austen 2015).

4.5 More Examples of AFIs

Yet more AFIs are covered in Table 4.2 spanning a wide range of sectors. The low-cost quadruped robot of entry 2, *Solo* shown in Fig. 4.3, is an AFI based on open source for software related to its control. Its creators, academics involved in cutting-edge robotics research, have contrived this AFI by using: 3D printed parts, off-the-shelf components and, of course, open-source control software (Grimminger et al. 2020).

The *Drugs for Neglected Diseases initiative* (DNDi) in entry 3 was kick started by *Médecins Sans Frontiéres* (or *Doctors without Borders*) for initially developing combination therapies and drugs for neglected diseases, which are typically not undertaken by major pharmaceutical or Big Pharma firms (Maxmen 2016). DNDi is also into developing commercial drugs, the domain of Big Pharma, from scratch at a fraction of the billions of dollars spent by these for-profit pharmaceutical firms (Maxmen 2016).

The next entry of *Impulse Drum Charger* (IDC) is a compact and intelligent contraption that functions as a low-cost turbocharger for inexpensive, small-displacement motorbikes (Cruccolini et al. 2018; BikeBandit 2018). The IDC is an inexpensive Bolton unit that comprises mechanical elements such as membranes, which keep its design simple. Developed by AlterEgo Hardware, the IDC is a good example of innovating under artificial constraints on resources through systematic application of S&T.

Entry number 5 concerning a *point of care* (POC) and hence a portable system for diagnosing equine respiratory infectious diseases is another example of an AFI based on mobile phony reported recently by Sun et al. (2020). A Motorola-Nexus-6 smartphone is coupled through a cradle to a microfluidic chip containing the pathogen samples (collected from equine nasal swabs) together with the dye for fluorescence. Such a system is inexpensive, fast and crucial for the remote collection of data on infectious pathogens for the timely mapping of their spread. Such an AFI has serious ramifications on controlling onslaughts such as the latest wrought about by the animal-based virus, SARS-CoV-2. Such AFIs will be very helpful from here onwards in fighting pandemics.

Another *point-of-care* smartphone-based AFI is the *transmission, reflection and intensity spectral* (TRI) analyzer of entry 6 developed by researchers at the *University of Illinois at Urbana-Champaign* (UIUC) (Long et al. 2017). The TRI analyzer, having multiple biosensing modalities, is a versatile POC system that handles a wide variety of commercial laboratory-based assays corresponding to a wide range of biological pathogens. It is in fact a portable diagnostic laboratory that handles a variety of spectrum analyses including colorimetric absorption, fluorescence emission and resonant reflection. All of this for just $550 a piece of the TRI!

The rotary Wankel engine listed as entry 7 is an AFI that is finding applications for powering aircrafts and automobiles (Weiss 2016; Vorraro et al. 2019). It is a rotary version of an *internal combustion engine* with a patented *Self-Pressurizing-Air Rotor Air Cooling System* (SPARCS) for further improving the efficiency of a rotary design (Weiss 2016; Vorraro et al. 2019). This engine has fewer moving parts, as opposed to the traditional and bulky piston type, and hence prone to lesser vibrations. Also, the rotary design has a high power to weight ratio which is a compact product and low in cost vis-à-vis piston type engines. Other than the Japanese automaker Mazda Motor Corporation, Advanced Innovative Engineering (AIE), a British company, is also at the forefront of real-time applications of this technology as seen in Fig. 4.4.

The concept underlying the plasma-based particle accelerator, SLAC, described in Table 4.1 has been adopted by the CERN, Europe's flagship particle-physics laboratory (Gibney 2015; AWAKE 2016). This program called the *Advanced Wakefield Experiment* (AWAKE) uses pulses of protons instead of electrons as at SLAC to accelerate charged particles. The AWAKE program, an AFI shown in Fig. 4.5, by making accelerators compact through plasma technology has like SLAC reduced the high costs typical of conventional types that are spread over several Kilometers.

In fact, there is now a research institute dedicated to just small-scale particle physics of the type in entry 8 and related innovations that make it possible. The *Center for Fundamental Physics at Low Energy* (CFP) setup at Northwestern Uni-

Fig. 4.4 A roadster powered by the low-cost rotary Wankel engine. Image is taken from Weiss (2016)

Fig. 4.5 The AWAKE program of CERN using frugal plasma technology in accelerators. Image is taken from AWAKE (2016)

versity, Evanston, Illinois, is one of a kind. Under the leadership of Professor Gerald Gabrielse, the CFP uses small-scale approaches to test the standard model of particle physics using sensitivity from precision instead of leviathan high-energy apparatus (Popkin 2018).

Entry 9 pertains to yet another achievement from the Indian space agency, ISRO. The space agency has managed to launch a second Moon mission, Chandrayaan-2, involving an orbiter, lander and rover for just $93 million (Padma 2017). Other than the low-cost MARS mission of Table 4.1, the Moon mission is yet another significant tribute to the Indian space agency for researching into making AFIs from such very sophisticated low-cost missions! A note here on the brilliance and ingenuity of India's space researchers. With its humble beginnings due to cost constraints, India's space program has lifted off to ever greater heights. Constraints on resources in terms of knowledge, capital and also materials, did not stop the scientists at ISRO from inventing amazing rockets and probes. The manifold difficulties of such endeavors, when compared to efforts by other nations require the best solutions that work despite these constraints. Figure 4.6 encapsulates aptly both the humble beginnings and brilliant research of the ingenious workforce at India's ISRO.

The next entry presents an AFI for *Magnetic Resonance Imaging* (MRI). A low-cost MRI has been developed by researchers at the *Harvard Medical School* using smaller magnets for reducing outlays (Nature 2015). Radio-frequency coils pick up the weaker radio signals generated by the tissue exposed to the fields of smaller magnets. These signals are processed at a fast pace to produce images showing the presence of any larger abnormalities. Although this AFI cannot compete with the wider functionality of conventional MRIs, it excels in detecting major tissue abnormalities in the shortest possible time vis-à-vis clunky counterparts at one-twentieth of their cost.

Entry 11 reports an AFI related to the technique of spectrometry. Traditional spectrometry finds varied applications in areas including biomedical, pharmaceuticals, energy, forensic science, environment and chemistry and chemical engineering. A portable, smartphone-based spectrometry using *colloidal quantum dots* (CQDs), shown in Fig. 4.7, has been reported that saves on weight and instrument complexity, and hence costs less than its traditional counterpart (Bao and Bawendi 2015).

Entry 12 pertains to satellites smaller and also cheaper than *CubeSats* listed in Table 4.1. These smaller versions, called *chipsats*, are literally like tiny circuit boards, depicted in Fig. 4.8, surviving the harsh environment of space (Jones 2016; Space 2019). These probes costing about $100 a piece are showing potential for many applications.

Yet another *point-of-care* smartphone-based AFI is used to detect the presence of BRCA-1 gene for breast cancer (Prasad et al. 2019). A 3D printed device called *FluoroZen*, shown in Fig. 4.9, takes in the sample for detecting fluorescent oligonucleotide spots on nitrocellulose (NC) paper, whose creation is rooted in microfluidics. These spots, representing the presence of BRCA-1, are shown on the screen of a coupled smartphone with an accompanying message confirming the presence or absence of cancer-related gene.

Fig. 4.6 India's brilliant space program offers some of the best solutions in space research. Image is taken from Twitter Feed of Prasar Bharati (https://twitter.com/prasarbharati/status/831786401620451330)

Fig. 4.7 Colorful colloidal quantum dots for low-cost spectrometry. Image is taken from Bao and Bawendi (2015)

Fig. 4.8 Smaller and cheaper than even *CubeSats*! Image is taken from Jones (2016)

The futuristic potential of AFIs in powertrain technology for internal combustion and hybrid vehicles is seen in Camcon Automotive's new *Intelligent Valve Technology* (IVT) (Stone et al. 2017; GreenCarCongress 2019). This technology does away with the traditional camshaft of traditional IC engines with precise digital control of valves through a set of electric actuators. Such precise digital control of combustion will significantly influence lowering of both *green house gas* emissions and costs. Currently setup in luxury vehicles such as Jaguars, IVT will make its debut as an AFI in mass market vehicles of the future. A picture showing this valve in a representative engine is shown in Fig. 4.10.

An interesting AFI shaping up seismology is again based on mobile phony. The accelerometers in the multitudes of ubiquitous smartphones around the world can function as cheap but good-quality seismometers. Based on the efforts of researchers at the University of California at Berkeley, who created the MyShake app, Google and Android operating systems, will integrate this earthquake-detection feature. Multitudes of such smartphones will beam relevant signals to Google to aid timely detection of earthquakes for corrective action (arsTECHICA 2020).

Entry 16 is associated with an interesting movement akin to the DIY for pollution sensors in Table 4.1 (again entry 16!) termed *open science hardware* (OSH). The OSH, analogous to "open software" in the computer-industry, has led to researchers building their own experimental DIY equipment (Gibney 2016). Sophisticated platforms for equipment including optical microscopy (FlyPi) and micropipette have been 3D printed to get significant cost reductions and also the necessary rigorous quality. An example of OSH developed by *Nature* and, shown in Fig. 4.11, involves a low-cost syringe pump, the software and hardware details being available on *Nature's*

Fig. 4.9 A 3D printed device called *FluoroZen* couples with smartphones to detect cancer affordably. Image is taken from Prasad et al. (2019)

Fig. 4.10 Camcon's *Intelligent Valve Engine* for internal combustion in vehicles. Image is taken from GreenCarCongress (2019)

website for building this device. Such an OSH syringe-pimp costs $100 as opposed to $250–$1500 for the traditional one (Gibney 2016).

Rolls Royce's LibertyWorks division based in Indianapolis, US has announced the successful testing on test rigs of small and affordable gas-turbine-engine cores (Norris 2020)—shown in Fig. 4.12. These jet-engine cores have been developed for attritable flights with an emphasis on drones and possibly missiles. A critical aspect of this development was the incorporation of low cost in the inception and design stages of the core by researchers working for the company. Low cost was mainly achieved by

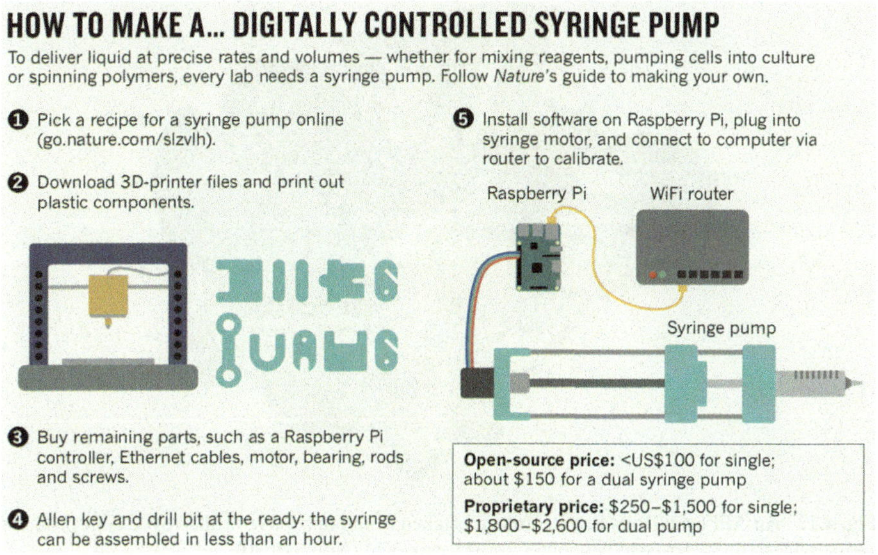

Fig. 4.11 DIY for scientists to make open-source AFIs! Image is taken from Gibney (2016)

using fewer parts and also *additive manufacturing* (AM) or 3D printing techniques. Moreover, the technology is scalable through its scalable design, manufacturing and materials techniques.

Entry 18 lists an interesting AFI that originated from a whirligig—that most of us have enjoyed playing with as children. Researchers headed by Professor M. Saad Bhamla from *Georgia Tech University*, US have developed a low-cost centrifuge called 3D Fuge for biomedical, clinical and other applications (Byagathvalli et al. 2019). As shown in Fig. 4.13, this is a low-cost AFI that is 3D printed and hand operated requiring no power source. These are ideally suited for remote and/or benchtop applications. These centrifuges have shown performance at par with standard benchtop types.

Gravitricity (https://gravitricity.com/) is the next AFI which uses earth's gravity for generating peak power outputs of between 1 and 20 MW (Irving 2020). Similar to *Gravitylight*^{TM} that uses gravity as an energy source, *Gravitricity* uses renewable energy sources to lift the 500–5,000 ton weights up to heights commensurate to required potential energy. These heavy weights are then "dropped" under gravity for generating power. One such application is shown in Fig. 4.14. *Gravitricity's* amazing AFI is still under development and has the hallmarks of a low-cost sophisticated *frugal* product developed by the keen scientific understanding of the workings of gravity.

Entry 20 showcases a most recent AFI based on paper-based electronic devices. These devices are being developed by Professor Ramses Martinez and his group at Purdue University, US (Medeiros et al. 2020). These paper- or cardboard-based

Fig. 4.12 An AFI from Rolls Royce. Image is taken from https://www.rolls-royce.com/products-and-services/civil-aerospace/helicopters/rr300.aspx/ and (Norris 2020)

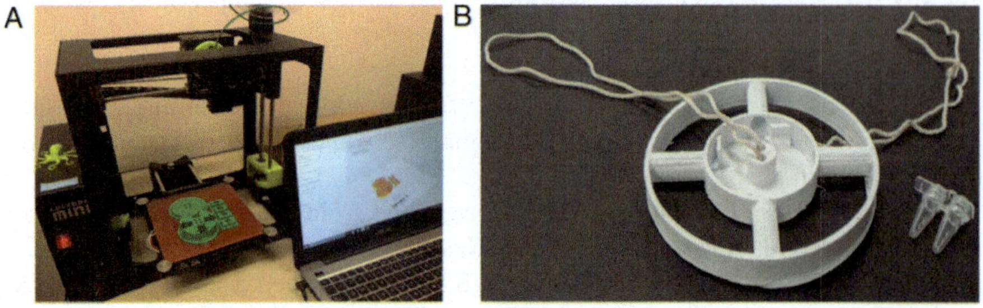

Fig. 4.13 Low-cost centrifuge from a 3D printed whirligig. Image is taken from Byagathvalli et al. (2019)

contraptions could be used as very low-cost human-machine interfaces including computer keyboards. Interestingly, these devices are powered by pressure exerted by the human subjects on the concerned paper or cardboard device. This AFI has been made possible by advances in flexible electronics and novel ways of printing these on moisture repellent paper.

Making low-cost medicines on the go is a great idea for an AFI. Professor Govind Rao of the *University of Maryland, Baltimore County* (UMBC) does just that with his portable briefcase like system called Bio-MOD (Arnold 2019) shown in Fig. 4.15. Bio-MOD stands for *Biologic Medications on Demand* and it can make advanced drugs for just a few dollars. Professor Rao's system is housed in a steel briefcase which is internally partitioned into a series of black boxes containing chemicals, etc., that are all controlled by an onboard computer to make cheap medication on the go. Other than medical emergency and defense applications, Bio-MOD like systems

Fig. 4.14 Raising loads under gravity to generate electricity. Image is taken from https://gravitricity.com/

can be a great *frugal* tool to deliver medication to the poor and/or needy, thereby enhancing cost-effective access to sophisticated drugs for all.

Entry 22 presents an interesting AFI in agriculture that has been termed as *frugal fertilizer* which does away with fertilizers by breeding plants for features appropriate for improving yields in nutrient starved soils (Gilbert 2016). This approach has been pioneered by Professor Jonathan Lynch of Pennsylvania State University in University Park, USA, who demonstrated the efficacy of this AFI in a variety of common beans called *Phaseolus Vulgaris* by breeding for root traits that facilitate uptake of phosphorus from soil with improved efficiency. Other than root traits including fine roots, lateral roots, basal roots, or appropriate root hairs, any other plant-based feature that enhances yield for deficiency in a given nutrient could be utilized. The absence of fertilizers not only reduces costs and makes it affordable to rich and poor farmers alike but this AFI also strengthens the case for *organic farming*. Moreover, such an AFI is a strong alternative to *genetically modified* (GM) plants whose use has been fraught with difficulties. Other than phosphorous and nitrogen, the *frugal fertilizer* approach has also shown improved yields for plants facing the onslaughts of droughts.

Another AFI harnessing the DIY movement is the LoCHAid developed by Professor Saad Bhamla (see Fig. 4.16) at the *Georgia Institute of Technology* in the US. LoCHAid is a really low-cost hearing aid developed from off-the-shelf components assembled together for giving a good hearing experience to people suffering from hearing loss (Lesté-Lasserre 2020). A match-box sized unit, hung around the neck

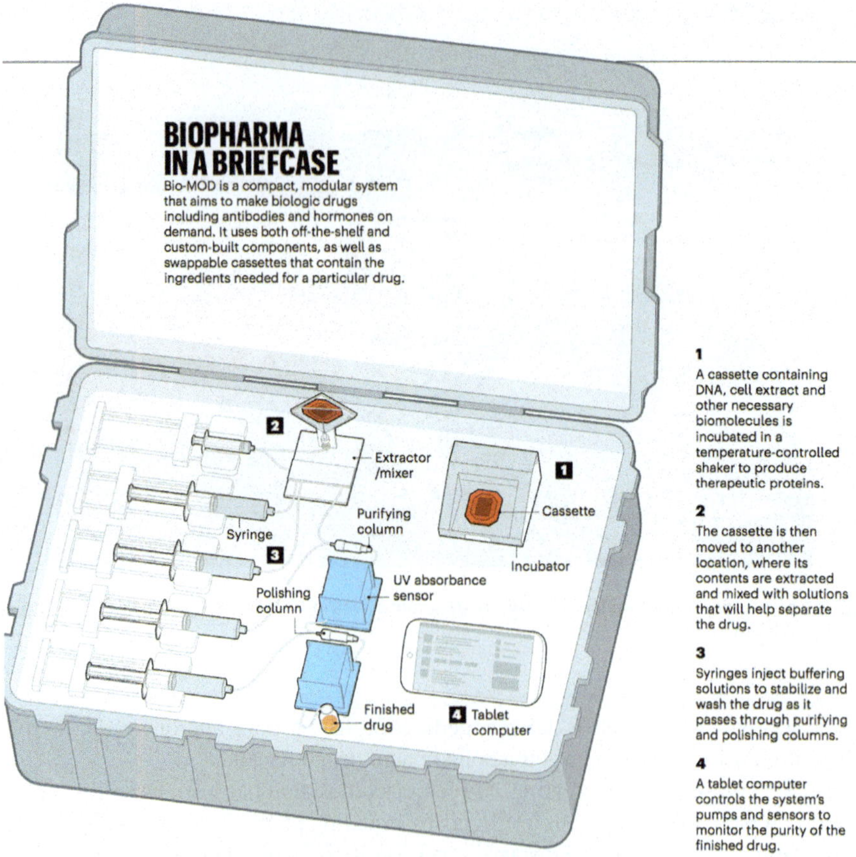

Fig. 4.15 Making medication on the go. Image is taken from Arnold (2019)

like a necklace, is made by soldering a microphone on a small circuit board with an amplifier and a frequency filter for sounds above 1000 Hz. This also comes with a battery holder, volume control, an on/off switch and a jack for plugging in earphones. Although LoCHAid sells for an amazingly low price of $1, blueprints can be downloaded and turned into actual devices for between $15 and $20.

A 17-year-old Maya Varma from California, US, has developed a low-cost *spirometer* for diagnosing lungs of patients with respiratory problems (Popular Mechanics 2015). Despite her age, Maya has used principles of electrical engineering together with 3D printing and her app savvy personality to develop this AFI which costs under $35. After the patient blows into the *spirometer*, which is a 3D printed Lilly pneumotachometer, the resulting lung capacity and flow rate are used together with relevant patient information by the app. This AFI can successfully diagnose with accuracy the following: COPD, asthma, emphysema, chronic bronchitis and lung disease.

Fig. 4.16 A $1 hearing aid developed by Professor Saad Bhamla at the *Georgia Institute of Technology* in the US. Image is taken from Lesté-Lasserre (2020)

A hand-operated washing machine called *Gentlewasher*, shown in Fig. 4.17, is an AFI because of the research efforts utilized in designing the flow-related features and other parts of this hand-cranked device (Shah 2018). The interior honeycomb surface of *Gentlewasher's* drum induces multidirectional flows that mix detergent and water with clothes effectively. Although the washing machine is small in volume holding about six pounds of clothes, it takes a mind-boggling 4–5 mins to complete the wash. The creators of *Gentlewasher* hope to reduce the current high price drastically with increasing sales. However, the compactness achieved from frugality in resource consumption effected through ingenuous engineering makes this an AFI that also delivers a good wash in a much shorter time.

Yet another example of radio astronomy-based AFI comes from retrofitted second-hand communications dishes that have been recently reported in another African country, Ghana (Wild 2017). As shown in Fig. 4.18, astronomers are retrofitting used telecommunications dishes that became obsolete after the coming of undersea cables. These retrofitted dishes, similar to entry 19 of Table 4.1, are part of the *African VLBI Network* (AVN), where VLBI stands for *Very Long Baseline Interferometry*.

Entry 27 presents a neat application of *carbon nanotubes* (CNTs), whose dimensions are of the same order as viruses, to trap viral particles (Yeh et al. 2020). The portable device called VIRRION (*virus capture with rapid Raman spectroscopy detection and identification*), shown in Fig. 4.19, uses Raman spectroscopy to identify the trapped viruses based on their vibrations. This inexpensive device gives reliable results in a few minutes, thus making it also suitable for POC applications.

AFIs have also invaded the turf of *ophthalmology*. Aravind Eye Hospital in Tamil Nadu, India, is a *frugal* champion when it comes to eye care (Popular Mechanics

Fig. 4.17 Hand washer that gives mind-boggling performance. Image is taken from https://echai.network/interviews/gentlewasher-the-washing-device-that-s-truly-gentle

Fig. 4.18 An old communications dish being retrofitted for telescopy in Ghana. Image is taken from Wild (2017)

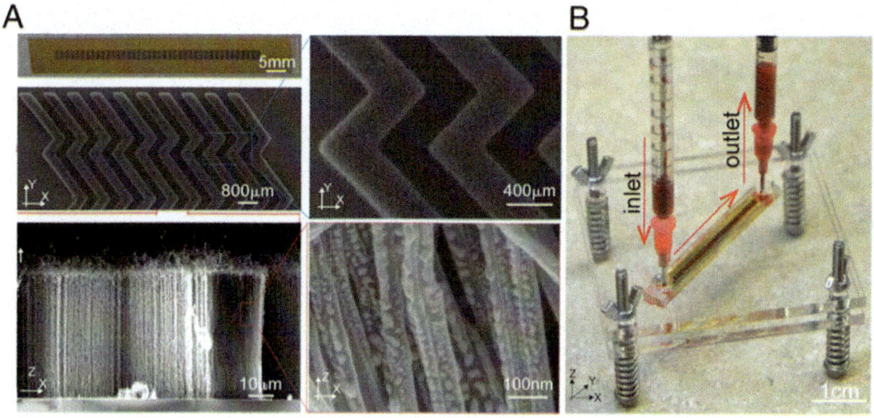

Fig. 4.19 *Carbon nano tubes* shown in an electron micrograph for trapping viral particles. The overall device, Virrion, as an AFI is also shown alongside. Image is taken from Yeh et al. (2020)

2016; Aurolab 2024). They have developed an *intraocular lens* (IOL) for treating patients with cataract for around $2 a pop. This endeavor was made possible through the joint efforts of the Seva Foundation and Aravind Eye Hospital in meticulously bringing the manufacturing plants for these devices, part by part, to India from the west. In fact, they are now doing extensive research into such AFIs for general healthcare and not limiting themselves to only eye care.

An interesting AFI from the Swedish furniture firm IKEA takes care of cramped spaces for living, especially in urban areas of the world (IKEA 2022). IKEA has created *multifunctional* furniture, a single unit which can have multiple functions. For instance, the company has created furniture nicknamed "Swiss" that combines a bed, sofa, dining space, and storage all in one unit. IKEA has created these and other types of units giving high-quality multifunctionality at low cost, thereby qualifying for being AFIs. Figure 4.20 shows the AFI, i.e., IKEA low-cost multifunctional furniture, which could possibly rope in robotics in the future (Lee 2019).

Another low-cost weather sensor is based on the strength of mobile phone signals. Such a sensor is based on the absorption and scattering of microwave signals by water and uses this to compare signals between cell towers to make predictions of precipitation at even street levels (Tollefson 2017). *ClimaCell*, a start-up in Boston, Massachusetts and many European and Israeli companies are pioneering this technique for global applications. It should be noted that a similar AFI collecting lightening data from sensors on cell towers to predict the onset of storms, etc., has also been covered in entry 17 of Table 4.1.

The Japanese entrepreneur, Mikitani Hiroshi, through his e-commerce firm Rakuten, is in the process of developing a low-cost architecture for 5G networks in Japan called OpenRAN (Economist 2020). Instead of relying on expensive hardware, Mr. Mikitani is making extensive use of software, inexpensive base stations and the cloud to develop this virtual architecture for the OpenRAN network. If suc-

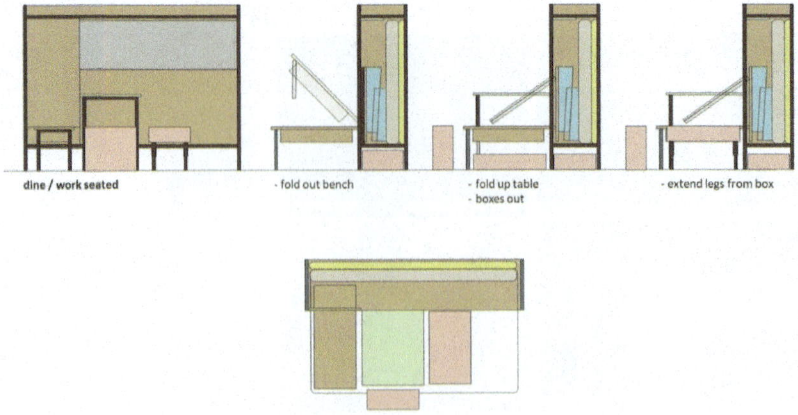

(a)

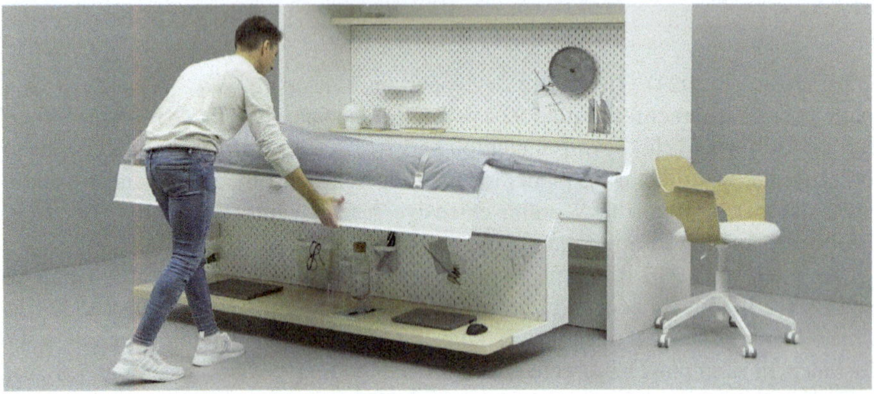

(b)

Fig. 4.20 Multifunctional furniture by IKEA as AFI. Image is taken from IKEA (2022)

cessful, OpenRAN will be a low-cost AFI at the heart of the internet, a virtual 5G network.

The *Redox flow batteries* (RFBs) are environmentally benign AFIs in the world of batteries (Janoschka et al. 2015). Batteries are especially important to the intermittent supply of renewables such as solar and wind energy. The RFBs offer a low-cost solution by using organic polymers for charge storage along with inexpensive dialysis membranes that separate the anode and cathode in an environmentally benign aqueous sodium chloride solution as electrolyte. These water- and polymer-based batteries have an energy density of 10 *watt hours per liter* and current densities up to 100 *milliamperes per square centimeter* with stable long-term cycling capability. In addition to their environmental credentials, these RFBs are affordable and also scalable.

A healthcare-related AFI aids in enhancing success rates of image-guided surgery of cancerous tissue (Garcia et al. 2018). This particular AFI is an image sensor, shown in Fig. 4.21, that is bio-inspired from the million-year-old architecture of the Morpho butterfly's compound eye. The sensor is multispectral where spectral tapetal filters have been integrated with photodetectors to give 1000 × higher sensitivity and 7 × better spatial co-registration accuracy vis-à-vis clinical imaging systems in vogue. Moreover, this small and compact sensor weighing just 20 g costs about ~$20 to manufacture, thereby making it affordable even in resource-crunched surgical rooms with inadequate lighting.

Some other DIY-based AFIs that are at the cutting edge of technology include a confocal microscope and an optical microscope that have been put to good use (Ravindran 2020). Confocal microscopes typically cost around $1 million and are used regularly in biomedical research on tissues, etc. Professor Gustavo Batista Menezes, of the *Federal University of Minas Gerais* in Belo Horizonte, who was cash strapped developed a DIY confocal microscope out of a $1 plexiglass stage and a $2 infrared lamp from a hardware store. This bare-bones confocal microscope was used to visvualize mice cells *in vivo*, which subsequently got his work on the cover of *Hepatology*, the flagship journal of his discipline. As for the optical microscope, *OpenFlexure* shown in Fig. 4.22, a 3D printable design has been developed for use in the UK by Dr. Richard Bowman, a physicist, at the *University of Bath*. This microscope is capable of attaining super resolution. However, the low-cost price tag of $262 has made it popular in resource-poor regions of *Tanzania*.

Entry 36 presents an ingenious AFI utilizing solar energy to generate saturated steam, under high-temperature and high-pressure conditions, for functioning as an *autoclave* to sterilize surgical instruments (Zhao et al. 2020). This AFI, shown in Fig. 4.23, developed by researchers at the *Massachusetts Institute of Technology* (MIT) utilizes an off-the-shelf low-cost tube-fin solar absorber that is coupled to optically transparent and also thermally insulating tiles made of silica aerogel to drastically reduce heat loss from the fin surface. It was utilized in real-time surgical sterlization tests in Mumbai for generating saturated steam at 100°C with an efficiency of 56%. Such performance is spectacular vis-à-vis those using expensive optical heat concentrators. Such great performance costs about $38.

Even *machine learning* (ML) has been utilized in healthcare for offering sophisticated diagnoses for some cancers (Esteva et al. 2017). A computer system was trained with 129,450 digital images of medically diagnosed specimens through deep learning to predict malignant lesions on skin. The system, after training, along with expert doctors were shown unseen images for diagnoses. The ML trained system performed at par or better than doctors and thus is an AFI that lowers cost for initial screening of skin cancers.

An AFI compensating for scotopic vision is a bio-inspired strategy to undertake low-light imaging for various applications (Liu et al. 2016). This AFI, shown in Fig. 4.24, uses an all-optical solution, based on the elephantnose fish eye, to develop a *biologically inspired photosensitivity enhancer* (BPE) comprising multitudes of microphotocollectors (μ-PCs). An individual μ-PC is typically enclosed by parabolic sidewalls coated with a reflecting material for amplifying light-intensity collected

Fig. 4.21 *Morpho* butterfly's compound eye architecture used for making an image sensor for guiding surgery of cancerous tissue. Image is taken from Garcia et al. (2018)

under low-light conditions. The BPEs—low cost and also consuming zero power—can be utilized for any application involving low-light imaging for enhancing photosensitivity.

The next AFI is a low-cost *gravimeter* that is very sensitive to local variations in the acceleration due to earth's gravity (g) on earth (Middlemiss et al. 2016). *Gravimeters* have several applications including sensing hydrocarbon reserves, building of magma prior to volcanic eruptions and, even subterranean tunnels. This AFI, shown in Fig. 4.25, is a *microelectromechanical system* (MEMS) possessing sensitivity of 40 μGal/Hz$^{0.5}$ and being just a few cubic centimeters in size. This MEMS device comprises two anti-flexures and also a cantilever-flexure etched out of silicon that constrains the motion of proof mass in one direction at low frequencies. This stable configuration under low frequencies makes the MEMS accelerometer into a *gravimeter* (Middlemiss et al. 2016). The cost of conventional *gravimeters* is $100,000 and is also bulky; weighing up to 150 kg. The MEMS-based *gravimeter* is compact and very low in cost!

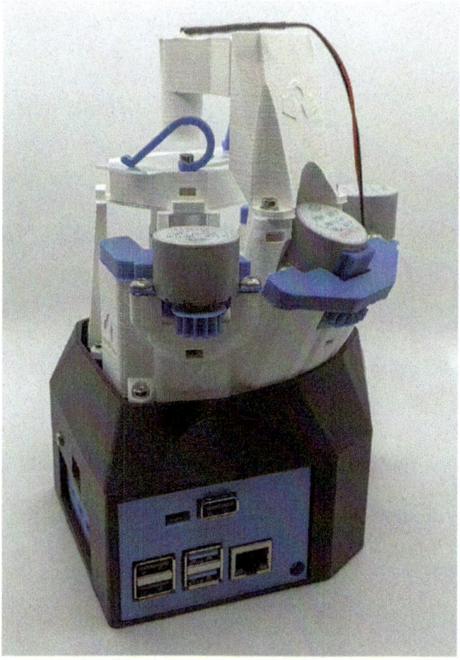

Fig. 4.22 A DIY microscope capable of super resolution optical microscopy. Image is taken from https://openflexure.org/projects/microscope/

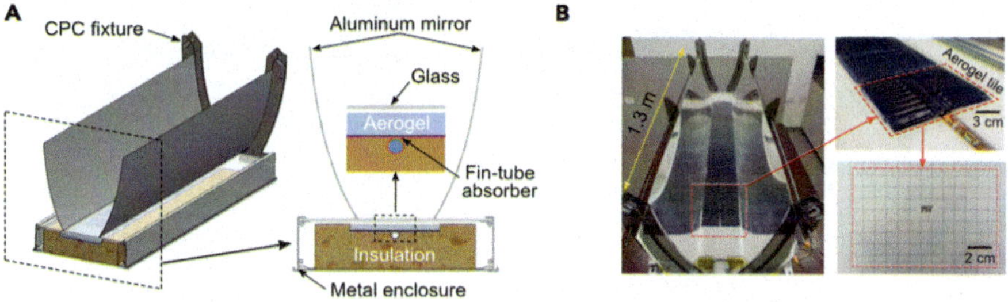

Fig. 4.23 Autoclaving through an AFI. Image is taken from Zhao et al. (2020)

OneWeb and Airbus is a company whose business is based on the mass production of an AFI (https://onewebsatellites.com/). The AFI here is a satellite for advanced space applications whose superior capabilities are offered at low cost. The cost-effectiveness is achieved through the mass production philosophy together with the modular nature of their satellites.

Entry 41 concerns a multi-locomotion millrobot called *Tribot* developed by researchers at the *Swiss Federal Institute of Technology Lausanne*, Switzerland, and *Osaka University*, Japan (Zhakypov et al. 2019). The simple assembly-free design of *Tribot*, shown in Fig. 4.26, is based on the locomotion mechanics of trap-jaw

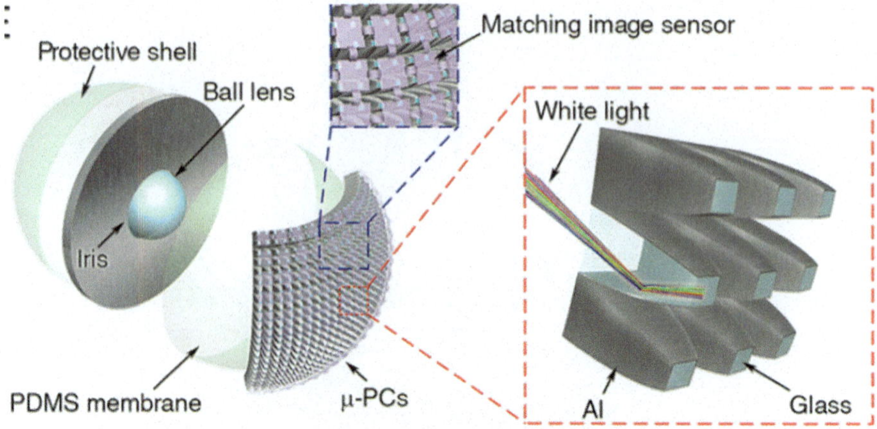

Fig. 4.24 A low-cost *photosensitivity enhancer* inspired from the eye of elephantnose fish. Image is taken from Liu et al. (2016)

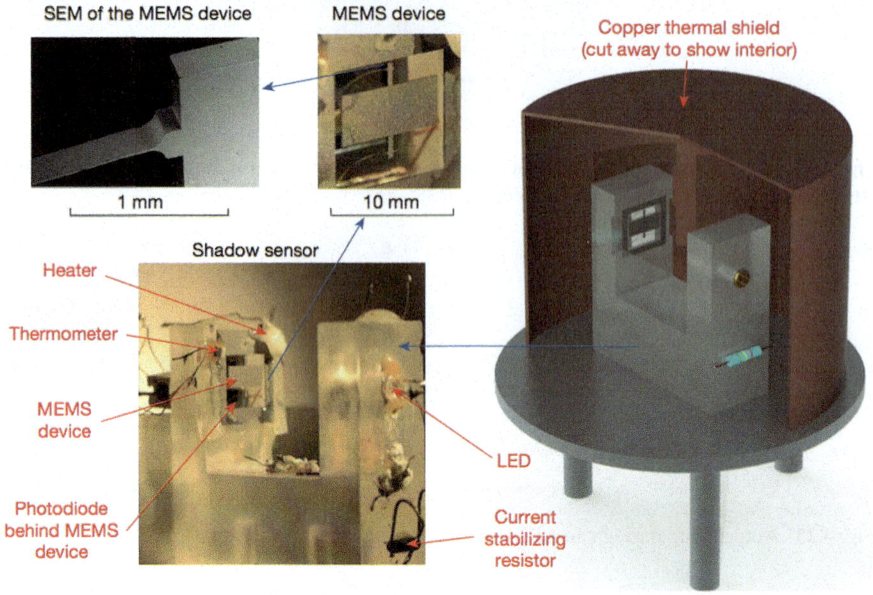

Fig. 4.25 A MEMS-based *gravimeter* as an AFI. Image is taken from Middlemiss et al. (2016)

ants that use their mandibles for moving on diverse terrains. The simple design is made possible by *printed circuit board* (PCB) technology and off-the-shelf parts that can be mass-produced without assembly. Simple design is a hallmark of any *frugal innovation*, and in the case of the *Tribot*, it also aids in assembly-free mass production. The *frugal* design together with mass production—another secondary tool to effect frugality—make this AFI cost just $60 a bot, and multitudes of such low-cost versatile *Tribots* can be utilized for various real-time collective tasks.

The advent of the *internet of things* (IoT) will connect billions of different types of devices around Earth into one very large communications-enabled network for not least, collective tackling of various tasks. Such a network has significant potential to lift the standard of living of society at large. However, cost is an important feature in the large-scale deployment of IoT, and an AFI coming from silicon *complementary metal oxide semiconductor* (CMOS) technology stands to make a big difference. CMOS-enabled nanostructured silicon harvesters are being designed for harvesting energy from heat dumped into the surroundings by a device as heat (Tarancón 2019). Such a harvester will do away with batteries and reduce the cost considerably while powering and also allowing continuous functioning of the IoT.

Yet another paper-based low-cost diagnostic for Zika virus has been developed by Professor James Collins of *Massachusetts Institute of Technology* and his colleagues (Pardee et al. 2016). A schematic of the working principle is shown in Fig. 4.27. These researchers modified an existing tool for Ebola virus that uses RNA sensors programmed and embedded in strips of paper. These strips get bonded to viral sequences leading to a reaction that changes the paper color. This technology was applied with modifications to detect the Zika virus with accompanying high sensitivity. The low-cost feature of this paper-based sensor with high sensitivity can significantly aid in fighting both epidemics and pandemics of Zika.

Dislocation of *patella* is a very common malady of the knee. The instability involved in the *patella* involves damage to the *medial patella femoral ligament* (MPFL). So restoring MPFL through surgeries avoids further damage to the *patella*. But existing techniques of locating *patella* to get to the exact MPFL site use X-ray imaging, which other than dangers of irradiation, is also expensive and prone to errors causing more problems. The concerned AFI is a device that locates and guides surgical tools to the *patella* which is crucial to MPFL associated surgeries (Dey et al. 2016). The AFI, shown in Fig. 4.28 in the form of a drill-guidance device, was fabricated using the low-cost 3D printing process. Successful cadaver-based trials have led to this AFI's utilization in MPFL reconstruction surgery. The cost of this device is about $20 a pop!

AFIs have also forayed into the depths of oceans. Entry 45 describes a Japanese endeavor to create low cost but very sophisticated spheres, shown in Fig. 4.29, equipped with technology to explore ocean depths for their environmental impact (Nature 2021). The low cost for these AFIs is made possible by manufacturing glass spheres possessing high sphericity coming from very precise finishing operations. These glass spheres are manufactured by *Okamoto Glass* for *Edokko Mark I*, which is the underwater observatory built by *research and development partnership for next generation technology of marine resources survey* (J-MARES). Such rigorous

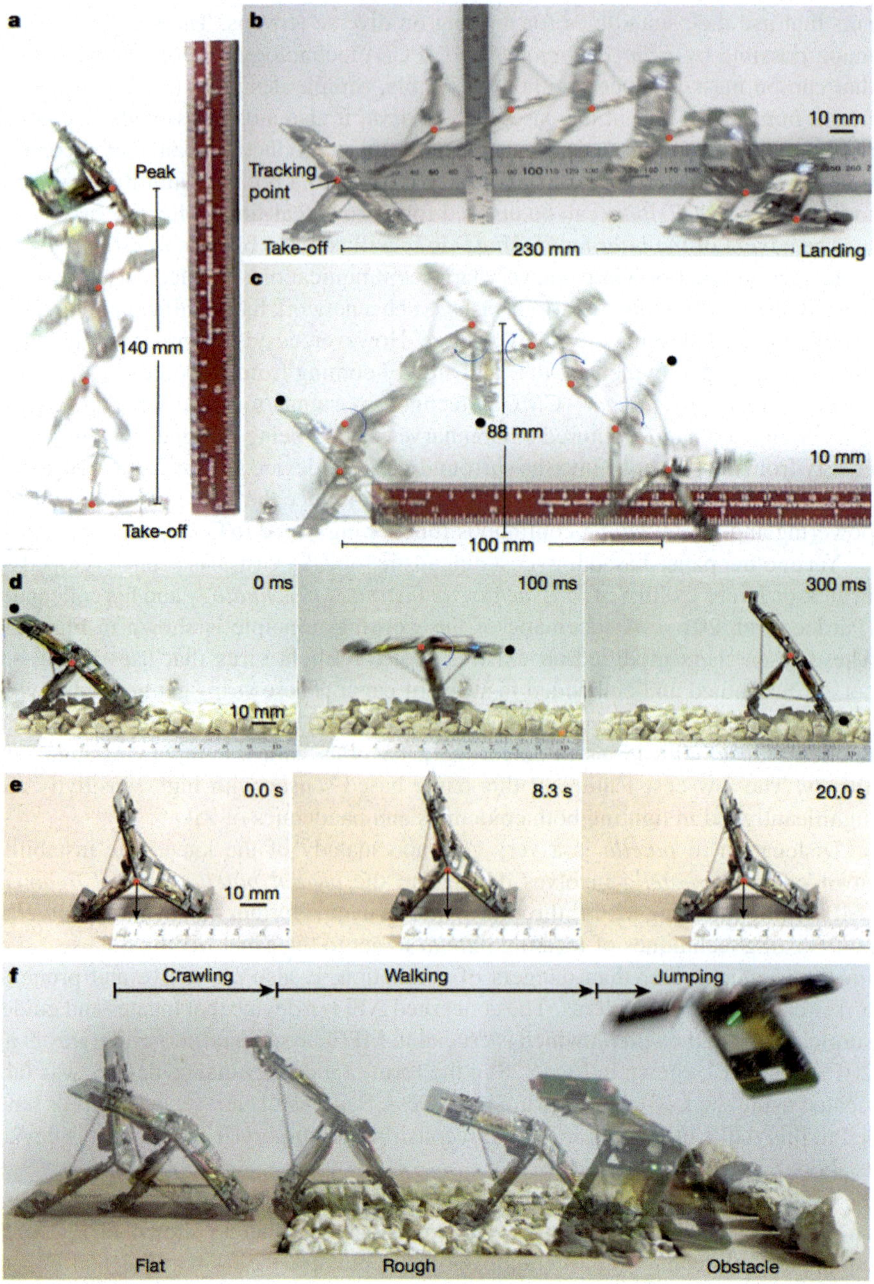

Fig. 4.26 An ant-inspired AFI called *tribot*. Image is taken from Zhakypov et al. (2019)

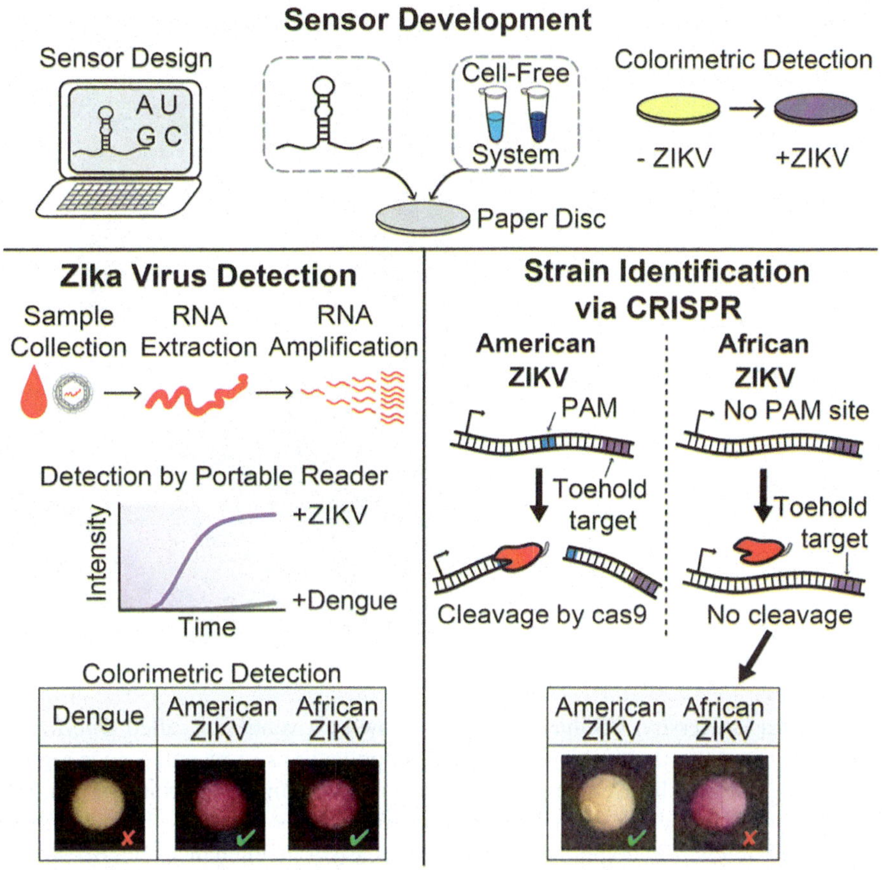

Fig. 4.27 Schematic of principle for low-cost paper-based diagnostic for Zika virus. Image is taken from Pardee et al. (2016)

design and fabrication, a recurring theme of AFIs, made the glass spheres resistant to ultra high-pressures of the deep ocean. The conventional alternative of using titanium alloy for the spheres is very expensive.

Sophisticated AFIs have also appeared in water filtration technology. An AFI has been developed to use the visible spectrum of sunlight for the photocatalytic disinfection of water (Liu et al. 2016). It uses vertically layered *molybdenum disulphide* (MoS2 or FLV-MoS2) films for harvesting visible light that is subsequently used to efficiently disinfect water. Moreover, the deposition of Cu and Au for facilitating electron-hole separation and also aiding the production of *reactive oxygen species* (ROS) reactions improved the ability of FLV-MoS2 to rapidly inactivate more than 99.999% of bacteria in a short time.

Entry 47 is a low-cost mechanical ventilator whose development predates the onset of the current Covid-19 pandemic (Stanford 2024). Dr. Matthew Callaghan

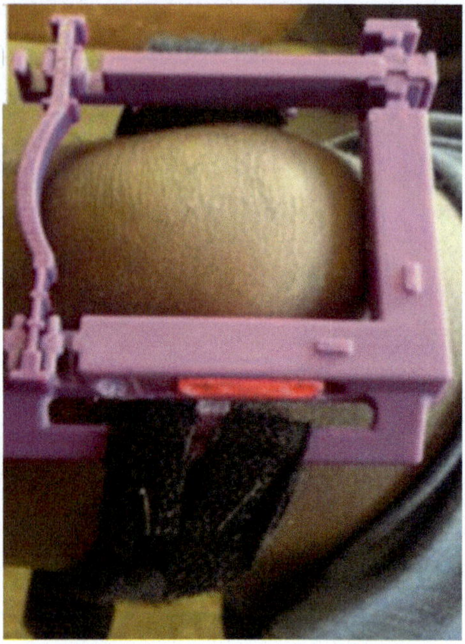

Fig. 4.28 AFI for guiding surgical tools to the *patella*. Image is taken from Dey et al. (2016)

of Stanford University, US, has developed a low-cost ventilator called *OneBreath*, shown in Fig. 4.30, that is proper for critical applications in resource-poor regions that can be operated by even novices. The design of *OneBreath*, a quintessential AFI, has accounted for nearly every handicap of resource-poor settings. It is built from scratch out of low-cost components and affords ease of maintenance. *OneBreath* can be used in both hospital and emergency-field sessions.

Affordable housing that is also stylish and catering to the sophisticated requirements of clients is the next AFI. Earl Patrick Forlales, who specialized in materials science and engineering at the *Ateneo de Manila University*, *Philippines*, created a low-cost housing idea called CUBO (BBC-News 2018). CUBO provides functional and sophisticated homes made primarily out of bamboo, a member of the grass family that grows abundantly in regions of the earth. The bamboo, which is heat treated and laminated, is resistant to earthquakes and emits more oxygen while not causing soil degradation during its growth and also harvesting. And CUBO offers various sophisticated designs by engineering eco-friendly bamboo at low cost. CUBO has even won the top prize for its AFI from the *Royal Institute of Chartered Surveyors'* (RICS) Cities for our future competition.

Entry 49 is a low cost and portable *optical coherence tomography* (OCT) scanner developed by researchers at Duke University, US (Song et al. 2019). OCTs are regular implements in the fight against macular degeneration and diabetic retinopathy. Conventional OCTs cost over $100,000, thus making the associated healthcare expensive. This AFI was made possible through a 3D printed spectrometer made

Fig. 4.29 A Japanese AFI in the form of high-sphericity spheres carrying technology to probe ocean depths. Image is taken from Nature (2021). This image has been provided by the *Development of Innovative Technologies for Exploration of Deep Sea Resources under framework* of the *Cross-ministerial Strategic Innovation Promotion Program (SIP) of Japan*

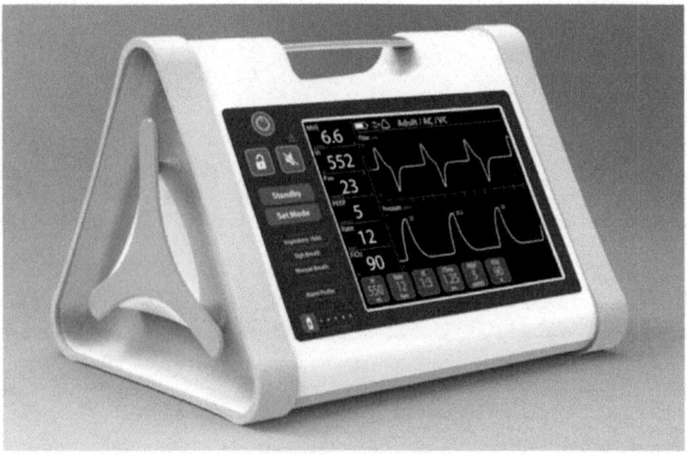

Fig. 4.30 An affordable mechanical ventilator made by a Stanford University researcher, especially for resource-poor regions. Image is taken from Stanford (2024)

of plastic, whose ingenious design, by Professor Adam Wax of Duke's biomedical engineering program, makes light travel in self-correcting circular paths as opposed to linear ones of traditional OCTs. Also, it is built from very low-cost parts, all of which makes this OCT cost just ∼$15,000 and also significantly lighter and portable than conventional clunky and expensive systems. Such low-cost high-quality OCTs can now be routinely used in resource-poor regions of the world. Figure 4.31 shows this OCT in action. This AFI shows the recurring feature of not sacrificing quality while embracing frugality!

This entry concerns with LEGO® brick system. There are many instances of LEGO® being utilized to create amusing toys including working robots for children and even adults! This AFI concerns use of LEGO® for making a functioning microscope, shown in Fig. 4.32, by a group of researchers at Münster University and Georg August University Göttingen, both in Germany (Vos et al. 2021). Other than the cost-effective LEGO® bricks and toolkit, this microscope used parts such as the lens of an Apple smartphone for high magnification; glass lens for low magnification; and acrylic lenses for ocular part. The total cost including LEGO® was €80, which is quite low by microscopy standards. This AFI was utilized to do some interesting experiments on imaging sodium chloride crystals; onion cells and Artemia swimming in water. This AFI brings out the need for workers with advanced skills, such as the German researchers, to make sophisticated low-cost applications out of mundane items.

Other than the many references to use of 3D printing or *additive manufacturing* in this chapter, AM deserves a separate "chapter" for its ability to create time- and cost-saving parts and also products. The DIY movement, also called "makers" movement, owes a lot of its successes to AM. For instance, OpenBCI, a company making affordable brain-computer interface, an AFI, uses 3D printing to fabricate

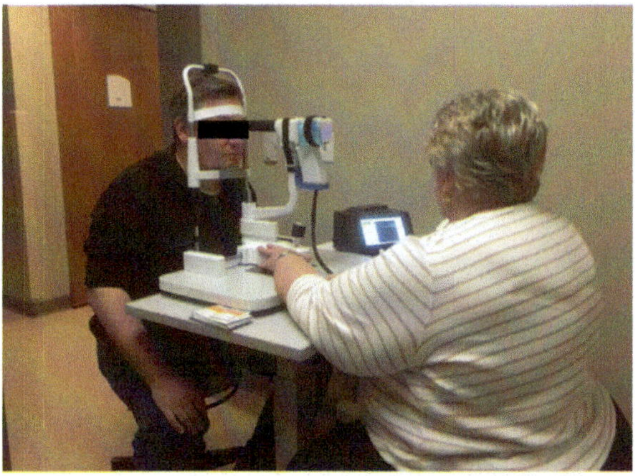

Fig. 4.31 The low-cost gun-shaped OCT used in imaging of a patient. Image is taken from Song et al. (2019)

a headset going like a skullcap (Scientific American 2016). This skullcap together with circuit board and open source software brings the technology of brain-wave research to even cash strapped but brilliant scientists, thus aiding the rapid progress of neuroscience. Another example concerns the use of AM to make affordable wind turbine blades, again AFIs. These blades, over 30 m long, cost about millions and are also time-consuming to make by the traditional way, but 3D printing has cut the time to weeks, instead of a year and, costs only thousands of dollars (Scientific American 2016). Hence, AM has been used by the *Oak Ridge National Laboratory* (ORNL), of the *Department of Energy* (DOE), in Tennessee, to make wind turbine blades that are also made efficient by printing different patterns at different points along the turbine blade! This variable feature is cumbersome to make the traditional way. Figure 4.33 shows the 3D printed mold and the resulting wind-turbine-blade.

Entry 52 highlights an *advanced frugal innovator* whose dedication to AFIs has won her the prestigious genius grant in 2016 from the MacArthur Foundation based in Chicago, Illinois, USA (https://www.macfound.org/fellows/class-of-2016/rebecca-richards-kortum). Professor Rebecca Richards-Kortum, of Rice University, USA, specializes in low-cost diagnostic techniques that could be used in resource-poor regions of the world. Of course, such AFIs then get applied in richer settings due to their superior performance. An example of one of her low-cost AFIs is a high-resolution microendoscope, shown in Fig. 4.34, that allows diagnosis and also the treatment of cervical cancer in a single go. Also, the application of novel contrast agents aids the sharpening of contrast in images of normal and precancerous tissue.

The next entry presents an interesting AFI in the form of a stand-alone "mini" drug factory shown in Fig. 4.35. Such a factory built out of "smaller" versions of the individual parts making up a conventional factory has been used in making a generic drug at CONTINUUS Pharmaceuticals, Massachusetts, USA (Nature 2019). Developed by Messrs Chuntian Hu and Salvatore Mascia, both researchers, the mini factory churns out an unnamed generic drug in flat 30 h. The purity is checked at

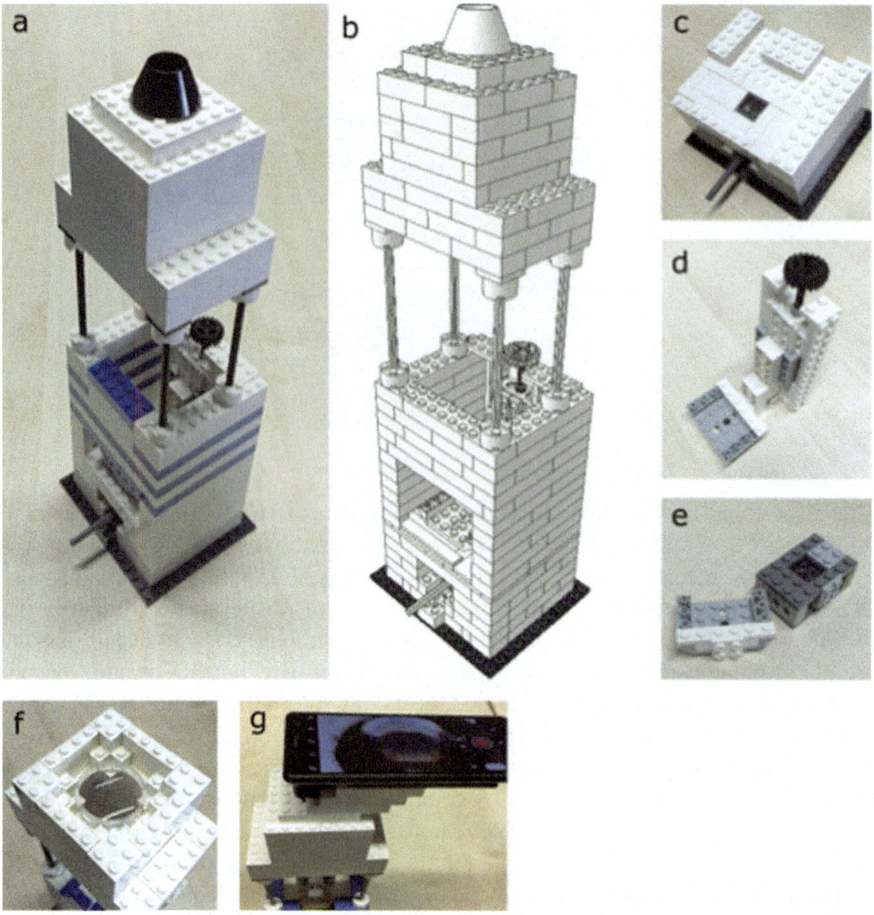

Fig. 4.32 A LEGO®-based optical microscope with off-the- shelf lenses. Image is taken from Vos et al. (2021)

various stages including the final one and this process is 35–40% cheaper and also uses only 40–50% of energy.

Entry 54 focuses on the prototype for an integrated laser accelerator developed by researchers at Stanford University, USA (Sapra et al. 2020). Traditionally, very expensive accelerators, such as SLAC *National Accelerator Laboratory* again at Stanford University, accelerate electrons over several kilometers to produce high-energy beams for several applications. In contrast, the new chip-based accelerator, shown in Fig. 4.36, imparts photon-based momentum from lasers to accelerate electrons. In particular, $\sim$0.9 *kilo electron volts* (keV) is imparted to some 80-keV electrons over a tiny length of 30 μm in a chip channel. The success of this idea will pave the way for relatively inexpensive miniaturized accelerators that will open up many new applications, not least in biomedicine.

Fig. 4.33 Two images showing **a** the *3D printed* mold used to make **b** the wind-turbine-blade. Image is taken from https://www.ornl.gov/blog/additive-manufacturing-wind-blade-molds-targets-time-cost-savings. Courtesy **a** of Oak Ridge National Laboratory, U.S. Dept. of Energy/Jason Richards and **b** of TPI Composites, Inc.

Stanford University is again referred to in this entry for a foldable optical microscope. Developed by Professor Manu Prakash and his team, this microscope called *Foldscope* blends principles of optics with the Japanese paper-folding technique of *Origami* (Cybulski et al. 2014). The resulting *Foldscope*, shown in Fig. 4.37, costs $\sim$\$1.0 and can perform imaging activities involving modalities such as brightfield, darkfield, fluorescence and lens array. The entire setup works through stretching and thumbing of an ingenious paper structure that houses the lens, LED light, etc. Moreover, it can be coupled to a smartphone to capture optical images. *Foldscope* is robust and can withstand the rough fieldwork of science and education.

A recent model of iPhone 12 Pro has builtin *light detection and ranging* LIDAR sensor thus making the camera capable of measuring depth (Leeming 2022). The iPhone achieves this depth-gaging ability by using a laser that sends 576 pulses in all directions. Although the company making the iPhone, Apple, introduced this feature for *augmented reality*, for Gregor Luetzenburg—Ph.D. fellow in Geosciences at the *University of Copenhagen*, Denmark—this feature enabled low cost and good quality measurement of erosion of cliffs. Figure 4.38 shows Gregor measuring erosion with

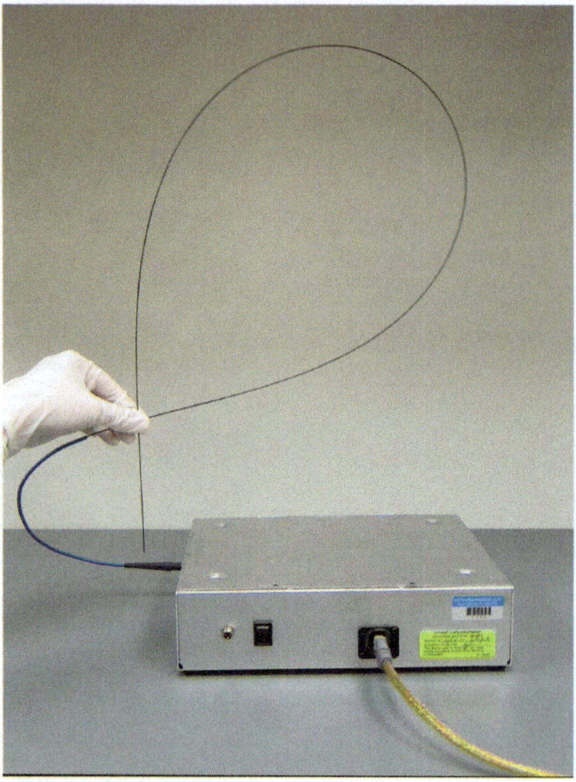

Fig. 4.34 A low-cost microendoscope with high resolution for diagnosis of various cancers. Image is taken from Quinn et al. (2012)

his iPhone. Again, conventional terrestrial laser scanners are big and also expensive and such an AFI infuses frugality into science!

Aluminum and its alloys are prized for their light weight, strength and recyclability. However, the traditional smelting of this metal is quite energy intensive and also expensive. Hence global aluminum synthesizers like Alcoa and Rio-Tinto are making this process low-carbon and also cheap. In particular, Alcoa together with Rio-Tinto, has through the *Elysis* venture, developed inert anodes plausibly of a ceramic composite to save significant amounts of energy dissipated during production and degradation of traditional carbon anodes according to a Nature-Editorial (2018). Moreover, Alcoa has even moved its operations from the US to Iceland, whose geothermal energy source is not only green, but also cheap. Overall, the combination of green anodes and green power can go a long way in eventually making aluminum production cheap and green (Nature-Editorial 2018)!

The next entry relates to a new technology that can be used to create novel AFIs. Structures can be built using *embodied logic* resulting in products that are self-actuating due to a combination of their geometry and composition. *Embodied logic*, which is a biomimetic technique, does away with power sources, sensors, electronics, etc., and the associated costs. The *venus flytrap* is an example of nature that is a self-

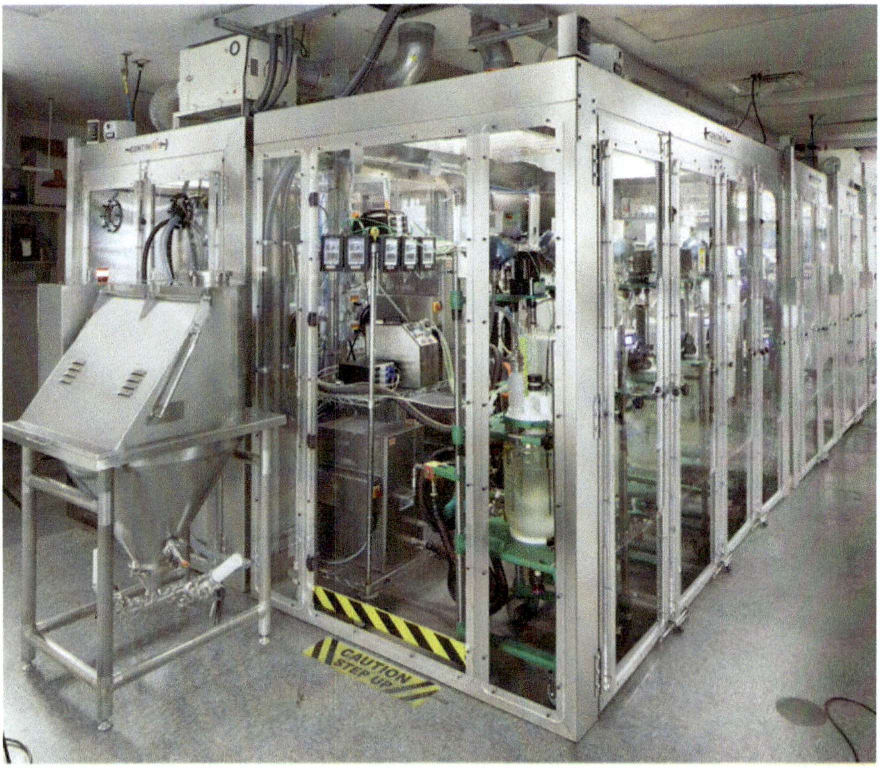

Fig. 4.35 By using smaller versions of conventional factories, this mini factory produces a generic drug at a significant discount in costs. Image is taken from Nature (2019)

actuating structure activated by the right stimuli. Jiang et al. (2019) used low-cost *3D printing* to create AFIs such as an artificial flytrap (shown in Fig. 4.39) and a hopper that flips on its own when unwanted chemical is imposed.

Minimally invasive surgery is popular due to, as the term suggests, minimal break in skin over the entire procedure. Robotic systems are used for such surgeries but they cost a couple of millions of dollars. Professor Shorya Awtar of the University of Michigan, Ann Arbor, USA, has developed a $500 human-powered system called FlexDex (Awtar et al. 2010), shown in Fig. 4.40. FlexDex is an AFI that mounts on a surgeon's arm and enables the instrument-tip to follow the movement of the surgeon's hand. It is an all-mechanical device that is ingeniously built for affordable surgeries around the world.

Additive manufacturing was used by *Continuous Composites*, based in the US, for printing wing spars for the *Low-Cost Attritable Aircraft* (LCAA) (Airforce-Technology 2021). The contract was awarded by *Lockheed Martin*, in the US, for building the main skeleton of the wing at low cost. Accordingly, *Continuous Composites* used their patented *Continuous Fiber 3D Printing* (CF3D) technology to

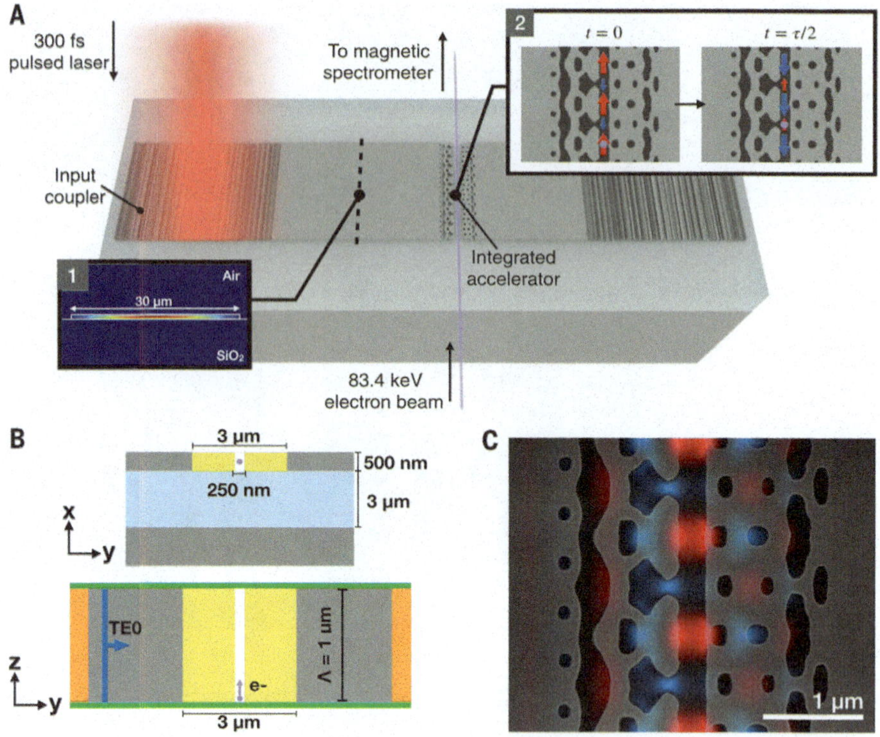

Fig. 4.36 A low-cost on-chip accelerator developed at Stanford University. Image is taken from Sapra et al. (2020)

additively manufacture the spar out of composites, shown in Fig. 4.41, by orienting structural fibers so as to resist highest loads. The resulting *3D Printed* spar could well resist loading in excess of design loads by 60%. Such performance is excellent considering fabrication without the tooling of traditional manufacturing of aerospace parts and their associated high costs.

Professors Terence Musho and Nigel Clark from the *West Virginia University*, US have designed and *additively manufactured* a *sparger* that aids efficient generation of geothermal energy at low cost (Miller 2021). The *3D Printed sparger head* generates bubbles, as seen in Fig. 4.42, that while rising through a geothermal fluid exchange momentum with it. This exchange causes the geothermal fluid to rise along with the rising air bubbles. The AM process reduces the cost of fabrication of the *sparger* for producing the required high-pressure air flow through the geothermal fluid. Besides low cost, AM or *3D Printing* was indispensable for the features required in the *sparger* that was designed for optimal geometry based on multi-phase fluid flow simulations. This novel method of raising geothermal fluids to generate energy at the surface is cost-effective when compared to traditional expensive methods for doing

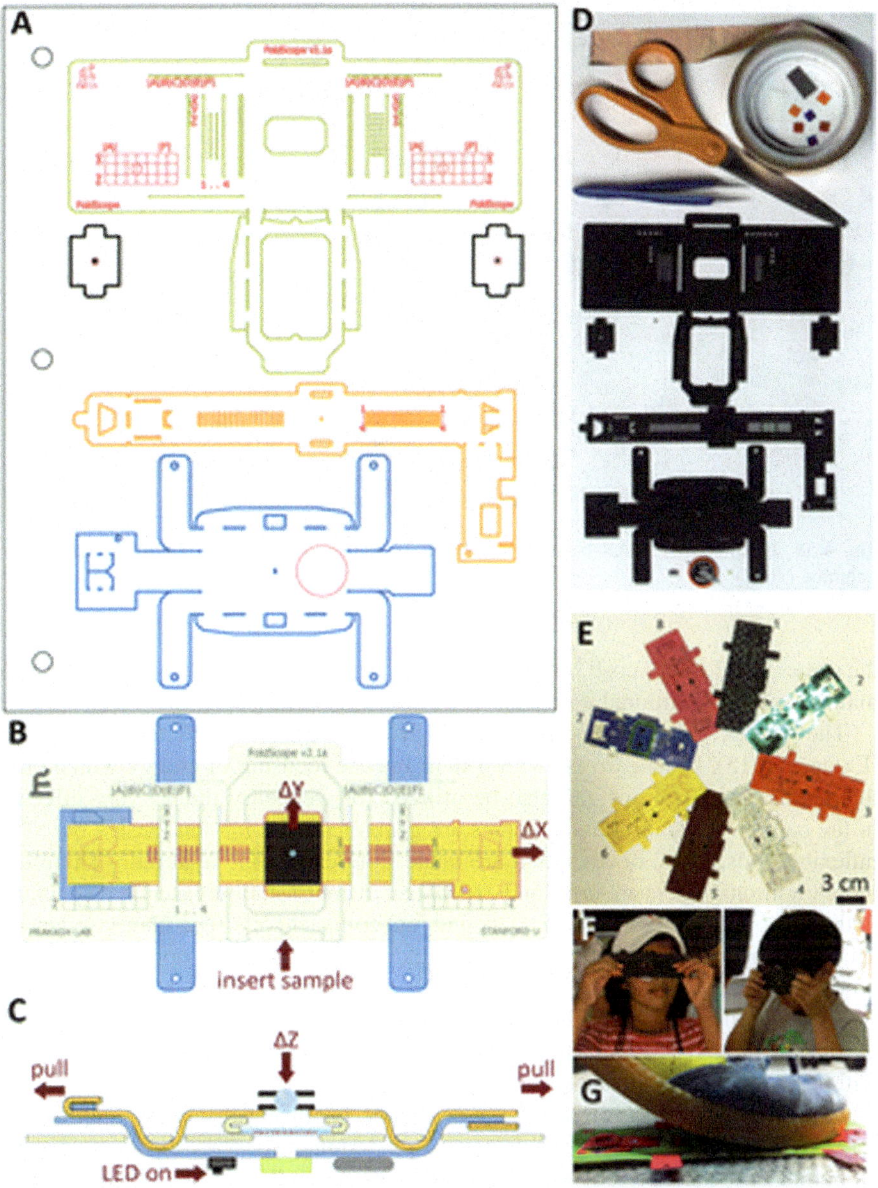

Fig. 4.37 A low-cost *origami*-based folding optical microscope. Image is taken from Cybulski et al. (2014)

Fig. 4.38 A low-cost LIDAR-based smartphone for measuring erosion. Image is taken from Leeming (2022)

the same. Besides low-cost *3D Printing*, costs are further reduced by the absence of maintenance of traditional pumps.

The next AFI is a weapon developed from cheaper but efficient fiber lasers (Extance 2015). Lasers are known for their intense focusing of energy which other than civilian applications can also be utilized for weaponry that is very effective in the battlefield. Traditionally, beams are generated by solid-state lasers based on inflexible rods, slabs, or discs that are made from crystalline materials. Therefore, these traditional lasers are large and expensive. In contrast, fiber lasers are generated in flexible optical fibers that can be wound into compact coils. The fiber laser and its accouterments are relatively inexpensive and have electrical-to-optical conversion efficiencies in excess of 30%. See Fig. 4.43 for a schematic of this AFI in weaponry. *Boeing* has utilized a 10 kW fiber laser for developing the *High Energy Laser Mobile Demonstrator* (HEL MD). The cheapest of traditional missiles are $100,000 a pop with *Boeing* HEL MD's one shot being less than $10!

A recent advance in telescopy for near-earth exoplanet discovery qualifies for a really hi-tech AFI. Exoplanet discovery predominantly relies on measuring the wobbling of a star due to a very small gravitational pull from the orbiting planet. However, a very precise version of such a technique is necessary for hunting exoplanets closer to our solar system such as around the star *Alpha Centauri*. This challenge was picked up by Professor Peter Tuthill, *University of Sydney*, Australia, to create a new type of telescope whose basic design is rooted in precision, low cost and agility (Tuthill et al. 2018; UnivofSydney 2021). *Telescope for Orbit Locus Interferometric Monitoring of our Astronomical Neighborhood* TOLIMAN is a low-cost telescope that is agile in its operation and uses a novel method using diffractive pupil lens. The

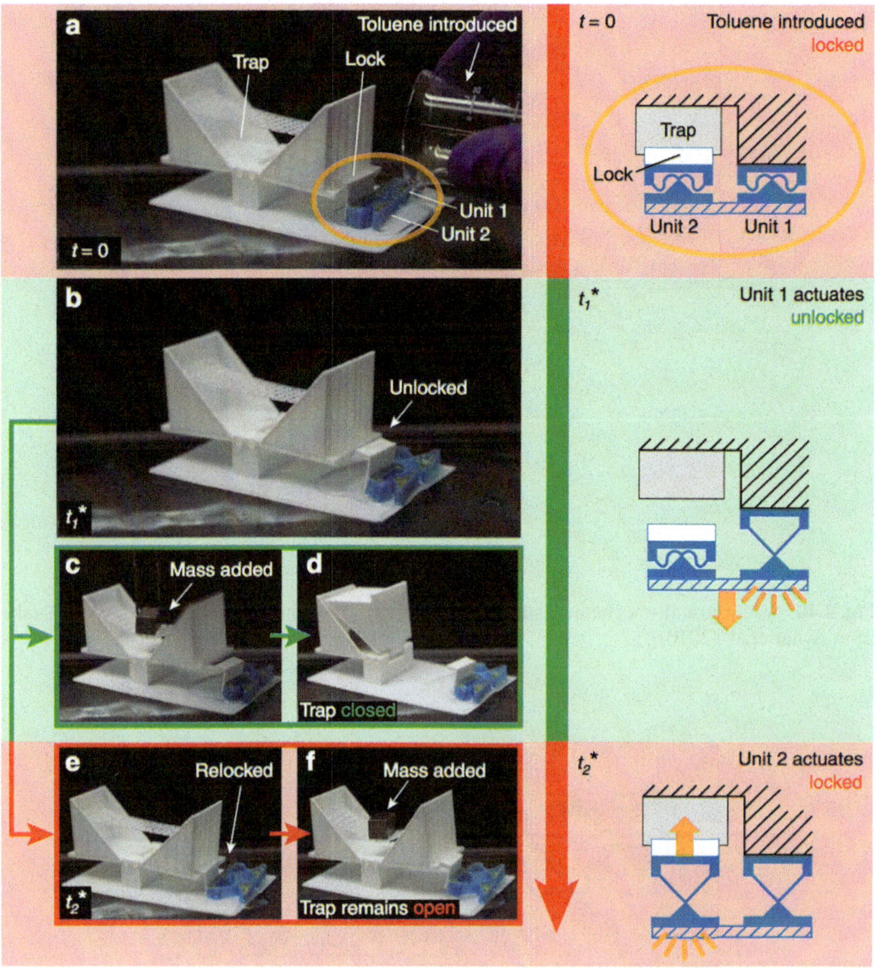

Fig. 4.39 A low-cost *embodied logic*-based self-actuating *flytrap*. Image is taken from Jiang et al. (2019)

mirror captures starlight into a complex flower type pattern which contains signatures of even tiny perturbations of a star being tugged by its smaller planets! Figure 4.44 shows the proposed design for Toliman.

The latest in solar cells qualifies for an interesting AFI. Professor Matthew Escarra at *Tulane University*, USA, and his co-workers have made a *Sunflower* pattern of solar cells and in doing that have used biomimetics in copying the effectiveness of this flower (Nature 2022). They concentrate their efficient solar cells at the center of the *Sunflower* and the waste heat generated is carried away by water which is heated to about 250 °C. This thermal energy could be converted to electricity through heat engines, etc. So this arrangement, of direct electricity and heat generation, gives

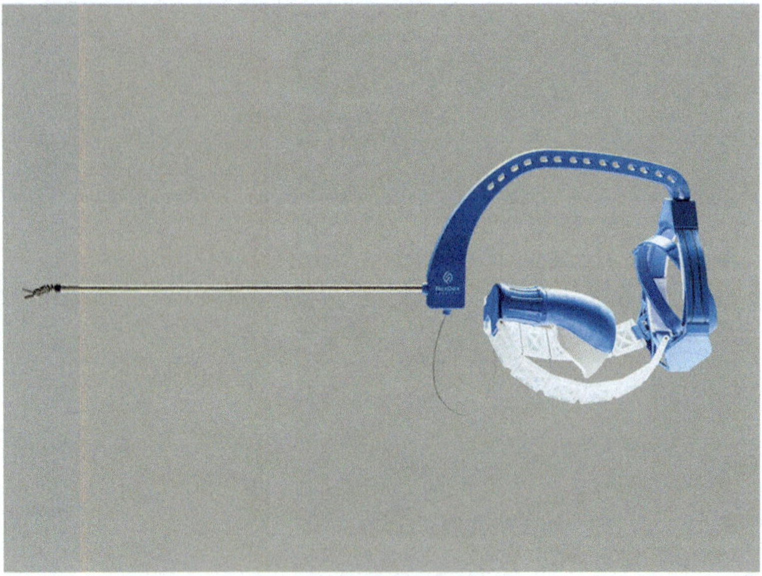

Fig. 4.40 A low cost all-mechanical surgical tool for minimally invasive surgeries. Image is taken from Awtar et al. (2010)

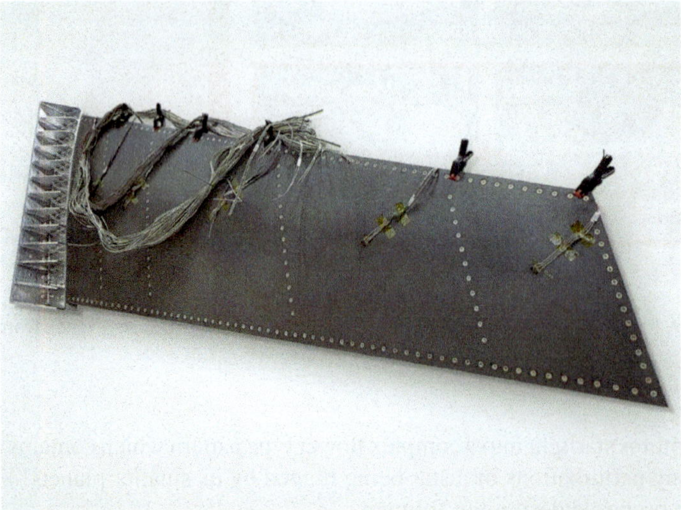

Fig. 4.41 A low-cost wing-spar printed by *Continuous Composites*. Image is taken from Airforce-Technology (2021)

Fig. 4.42 A low-cost *3D Printed sparger* for blowing bubbles through geothermal fluids for generating energy. Image is taken from Miller (2021). Courtesy of WVU Photo/JP Nesbit

an efficiency of 65%. Such quality performance beyond 50% efficiency of hybrid models has been achieved by a simple *Sunflower* design that is also inexpensive when compared to hybrid models. The *Sunflower* system is shown in Fig. 4.45.

The next AFI is a novel propulsion system for space applications. Electric propulsion has gained popularity due to its efficient usage of propellants. In particular, ion acceleration for generating thrust is typically generated through electron-induced ionization of the gas Xenon. However, Xenon is expensive; needs high-pressure equipment (again expensive) and is also in demand for other applications. Enter iodine! Iodine is abundant; can be stored as a pressure-free solid; and has better ionization properties, all of which make an iodine-based propulsion system relatively cheap with better thrust-to-power ratio. Researchers from France working for *ThrustMe* have developed an iodine propulsion system that gives superior performance when compared to those using Xenon (Rafalskyi et al. 2021). See Fig. 4.46 showing the iodine propulsion system. In addition to lower costs coming from the iodine-based propulsion itself, additional savings in materials and costs can be realized from simpler and compact satellites that will harness this new propulsion.

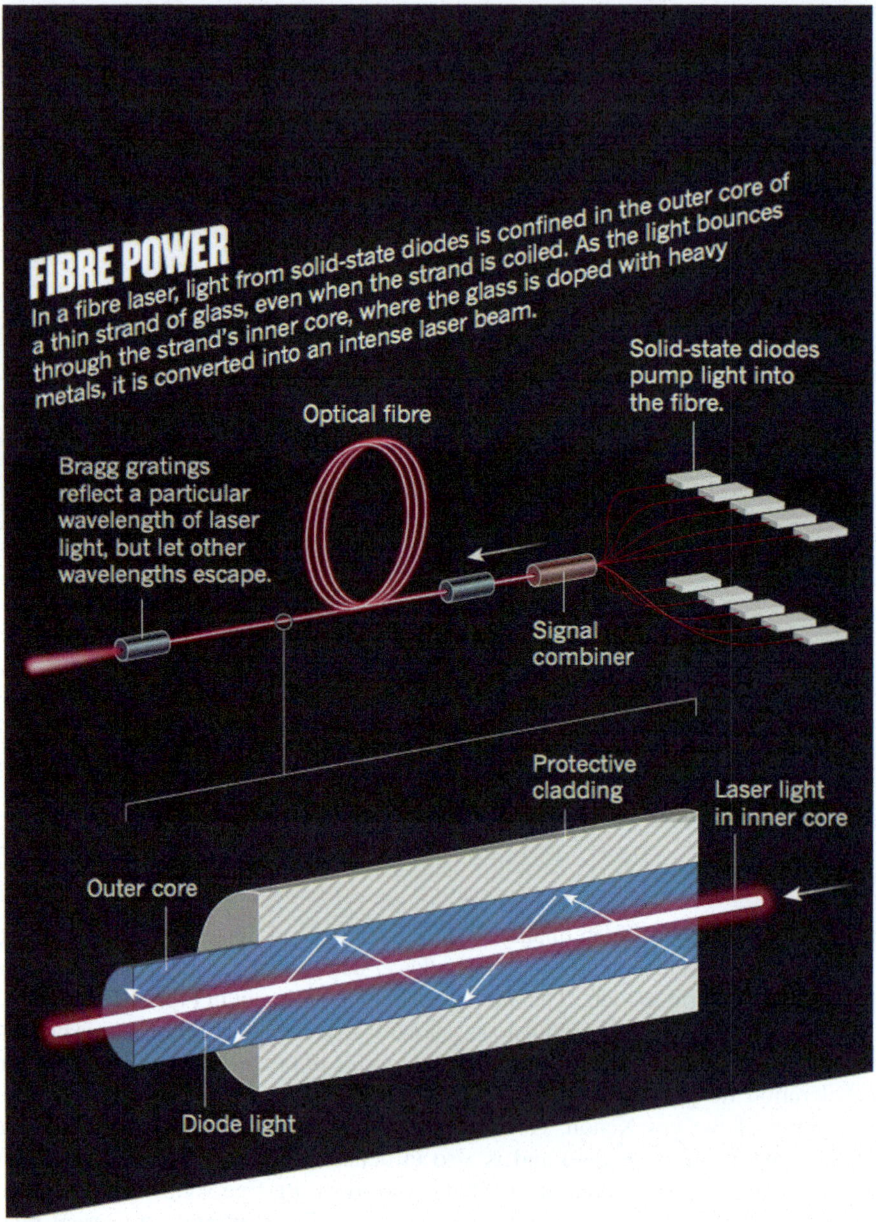

Fig. 4.43 A low-cost weaponry built out of cheap but very effective fiber laser. Image is taken from Extance (2015)

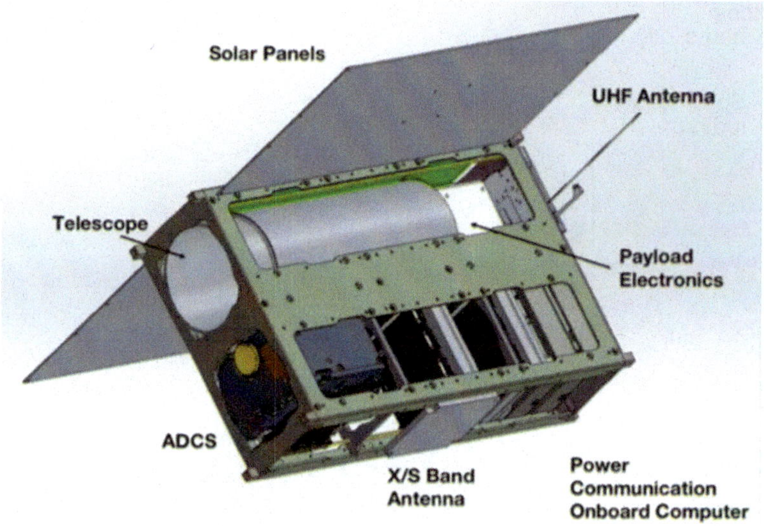

Fig. 4.44 Telescope built for extra precision location of exoplanets near to our *solar system*. Image is taken from Tuthill et al. (2018), UnivofSydney (2021)

Fig. 4.45 An efficient and affordable arrangement of solar cells in a *Sunflower* pattern. Image is taken from Nature (2022)

Japanese researchers have created an AFI in the form of a detector for studying Gamma-ray bursts released by thunderstorms on Earth (Gibney 2021), see Fig. 4.47. Such bursts typically associated with black holes and other cosmic phenomena have also been seen in thunderstorms on Earth. Their timely detection has serious ramifications for aviation and also terrestrial property in terms of plausibly avoiding harmful doses of radiation. These table-top detectors called *Compact Gamma-ray Monitor* (CoGaMo) are cost-effective and portable so that they can be affordably implemented

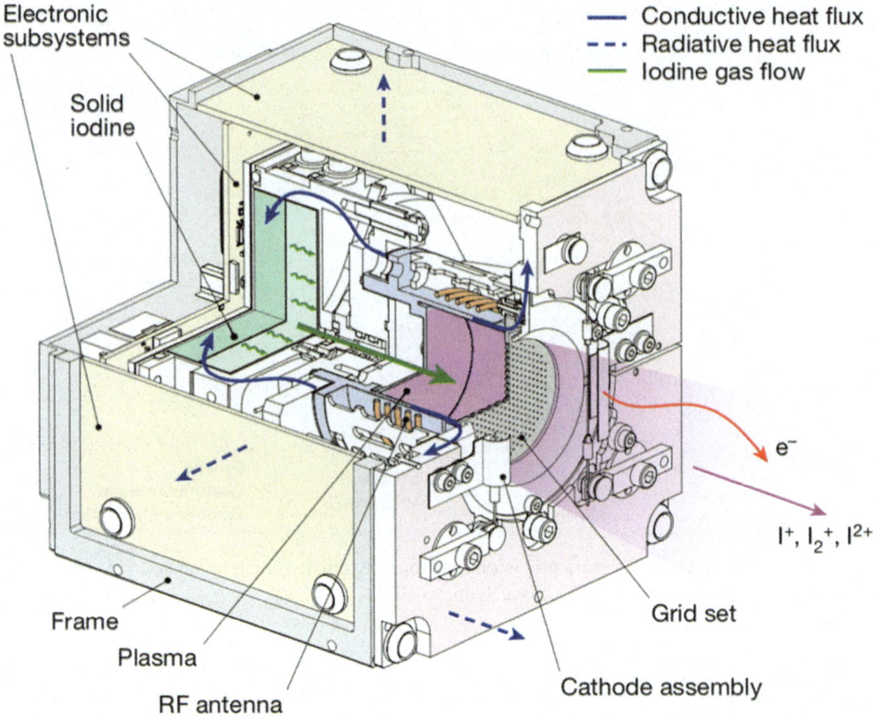

Fig. 4.46 Low-cost iodine-based propulsion for space applications. Image is taken from Rafalskyi et al. (2021)

through citizens. The CoGaMo itself is a GPS-based version of an earlier detector comprising a $60 Raspberry Pi computer and crystals of bismuth-germanate, which glow when exposed to Gamma-ray photons. Studies based on measurements from this AFI will throw light on the origins of such thunderstorm-based Gamma-ray bursts and subsequently enable the use of these detectors to forewarn possible radiation exposure, through lightning, on humans both on airplanes and land.

3D Printing has been utilized to create three-dimensional soft robotic actuators with sensors out of biodegradable materials. Researchers from *Johannes Kepler University* in *Austria* have developed a *fused deposition technique* for depositing biodegradable gels to create 3D actuators with integrated sensor networks (Heiden et al. 2022). Such conflation of sensor and actuator allows precise sensing of curvature, direction and force. This combination of *3D Printing* and biodegradable material takes lesser production steps compared to traditional molding and also uses lesser materials due to the involvement of additive manufacturing. Other than reduction in costs due to these factors, the creation of such soft robots is also environmentally friendly because the *end-of-life* actuator can be disposed of in water in which it readily dissolves. As seen in Fig. 4.48, the soft actuators are very effective in recognizing and also tackling obstacles. The development of this AFI is also an example of fru-

Fig. 4.47 An AFI detects radioactive Gamma-ray bursts from thunderstorms on Earth. Image is taken from Gibney (2021)

gality infusing into manufacturing and materials that are critical pillars of product development.

Some *point-of-care devices* have already been described in this section. Recent work by Professor Alexander Brem of the *University of Stuttgart*, Germany, and co-workers sheds light on *point-of-care devices* being inherently AFIs (Miesler et al. 2020). They give several examples of POCDs as AFIs for pregnancy test and diagnosing diseases such as influenza, malaria, ebola, tuberculosis and even HIV. Although the underlying tests use chemical and bio reactions, the output is visually given and also monitored by the ubiquitous *smartphone*. POCDs are used directly on the patient's bed; they reduce time significantly; they avoid idle treatments; offer measures against epidemics/pandemics; and are cost-effective. These features gel well with the definition of AFIs, and hence POCDs are inherently *frugal*. These inherent AFI features are depicted in Fig. 4.49.

Entry 69 concerns a new concept of mobility on the roads. Researchers at *Eli Electric Vehicles* have created a small and compact *electric vehicle* (EV) for personal

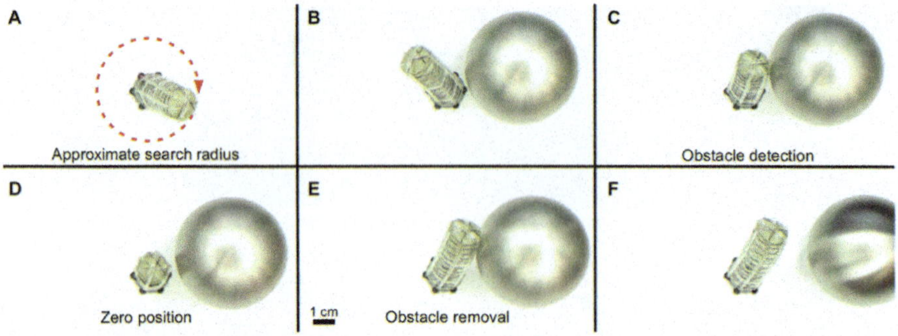

Fig. 4.48 An AFI in the form of a soft robotic actuator that is *additively manufactured* with biodegradable gel. This actuator affordably and effectively tackles obstacles in its path! Image is taken from Heiden et al. (2022)

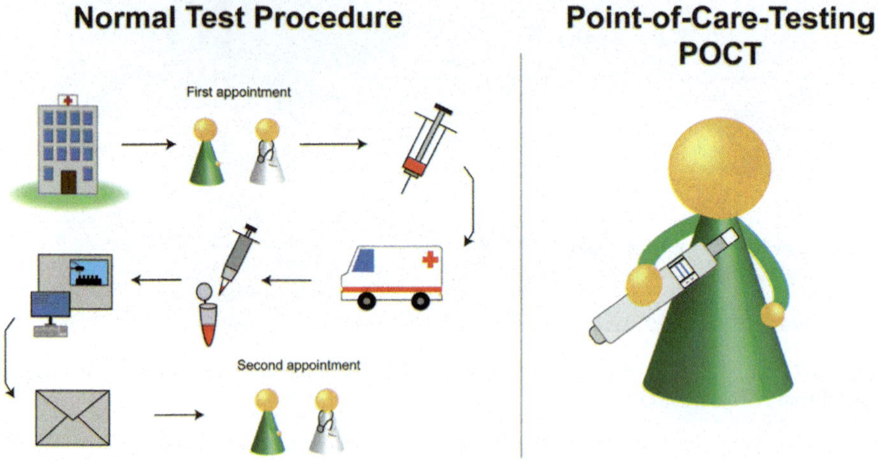

Fig. 4.49 *Point-of-care devices* are inherently *advanced frugal innovations* as depicted here. Image is taken from Miesler et al. (2020)

mobility. Another name for it is a micro-vehicle. The EV, shown in Fig. 4.50, is called ZERO and is made out of polypropylene—that can be reused—and aluminum to give a body that is light and durable. ZERO's Lithium batteries pack enough energy to give a range of 70 min and at 25 mph the ZERO can negotiate on both streets and roads. ZERO uses regenerative braking to generate energy from braking activity and its batteries are heated in winters to improve their efficiency. Although the speed seems low, it has been arrived at after conducting research into city driving. It is loaded with all kinds of goodies including sunroof, rear-view cameras, USB, etc., to make a comfortable drive. The current price at $16,000 (techcrunch 2021) is at the higher end for micro-vehicles but future versions should cost lesser as production begins and research further streamlines the design. *Eli's* ZERO is an AFI that partly

Fig. 4.50 A new concept using low-cost *micro electric vehicles* for personal mobility. Image is taken from techcrunch (2021)

uses salvaged material for its body and also squeezes out as much juice as possible from its power systems to give quality functionality affordably. This is a new concept some or all of whose features might feed majorly into the automotive industry.

An interesting AFI has utilized Wood for the construction of satellites. Yes, wood and not nickel-based superalloys like Inconel and other exotic materials have been utilized by Sumitomo Forestry in Japan (Economist 2021). Researchers there have utilized a composite made of lignin and cellulose that is highly resistant to tension and compressive loading. The material cost is low and is abundant with ease of manufacturing through chemicals. Plus, its manufacturing can be automated with more scope for lowering costs. Also, this satellite affords a simple design because all the communications antennae and other components dependent on radio waves can be housed inside the satellite "box" as wood is transparent to radio waves. Other than using a secret sauce for protecting the low-cost satellite from temperature extremes of space, such a satellite also readily burns off in re-entering the earth's atmosphere, thereby making disposal easier. This is an example of an AFI that lowers costs through cheap and abundant Wood which results in an affordable simple satellite design with cost-effective disposal on re-entry.

Entry 71 concerns an AFI for smart buildings that regulate energy consumption on their own (Kay et al. 2022). This is typically done by smart-windows the expensive versions of which use chemical reactions to shade themselves in times of need. Professor Benjamin Hatton and co-worker from the *University of Toronto*, Canada, have devised a pigment solution-based system that pumps the solution in a window to create shading from sunlight, see Fig. 4.51. Their AFI saves 30% energy at low cost vis-à-vis the chemical reaction-based window mentioned above. Their AFI is also an example of *Biomimetics* wherein they have used the mechanism used by ocean creatures called *Krill* in shading themselves from excessive sunlight.

The next AFI describes the latest example of Tesla latching on to the frugal movement (Arenaev 2023). Researchers at Tesla have created a new heat pump, as shown in a representative picture in Fig. 4.52, which is used to heat the cabin during winters and is a must in countries experiencing intense winters. A heat pump effect is air-conditioning carried out in reverse for heating applications. Researchers at Tesla have created a compact heat pump which is a simple single unit that does away with 15–20 component designs. Internal combustion engines are blessed with exhaust

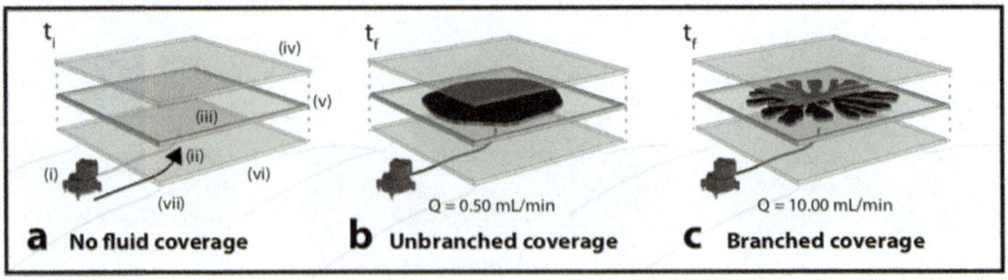

(a)

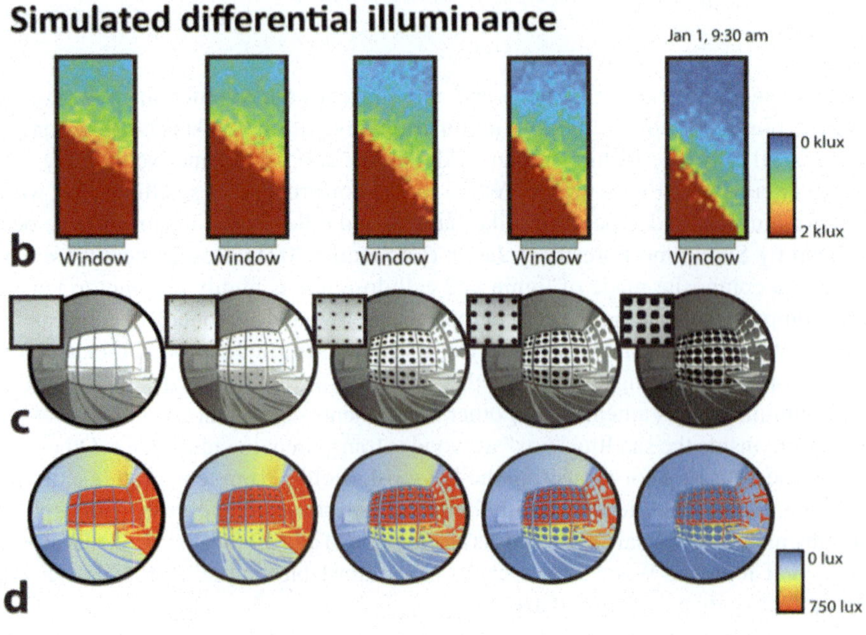

(b)

Fig. 4.51 Two images showing **a** smart window comprising pump and branching pigment solution and **b** working of the window showing gradual shading from sunlight. Image is taken from Kay et al. (2022)

systems that are nicely put to use in heating. However, electric vehicles do not have this luxury and have to typically develop intricate designs that are not compact, heavy and raise the cost. In contrast, Tesla's heat pump called Super Manifold has coolant and refrigerant channels in a single unit and which ingenuously tackles the heating problem under the worst winter conditions. Moreover, Tesla's heat pump is energy efficient!

Entry 73 deals with miniaturization in the field of optical microscopy. In particular, the AFI in the form of a miniature microscope, of 8 mm width, see Fig. 4.53, was assembled out of only five 3D printed or additively manufactured parts and some

Fig. 4.52 An AFI from Tesla in the form of a compact heat pump! Image is taken from Electrek (2020)

off-the-shelf components (Hai et al. 2023). A single microscope can be built at an eye-watering low cost of US $4! Such advanced frugality is made possible by the design of the microscope which allows the various parts to snap into proper alignment that traditionally requires costly skilled labor and expensive equipment. The design therefore allows quick fabrication and assembly in that the significant optomechanical parts are 3D printed in just 50 min!

The next entry highlights the success of Cuba, a country which despite sanctions has been punching above its weight in research. Cuba has set up several centers such as the *Center of Molecular Immunology* (CIM), *Center for Genetic Engineering and Biotechnology* (CIGB) and the *Center for Immunoassays* (CIE) that have produced world-class vaccines, and treatments for cancer and other disorders at low cost. All of these AFIs have been produced by workers with advanced skills including PhDs who utilize their *frugal* mindset for achieving these *advanced frugal* products.

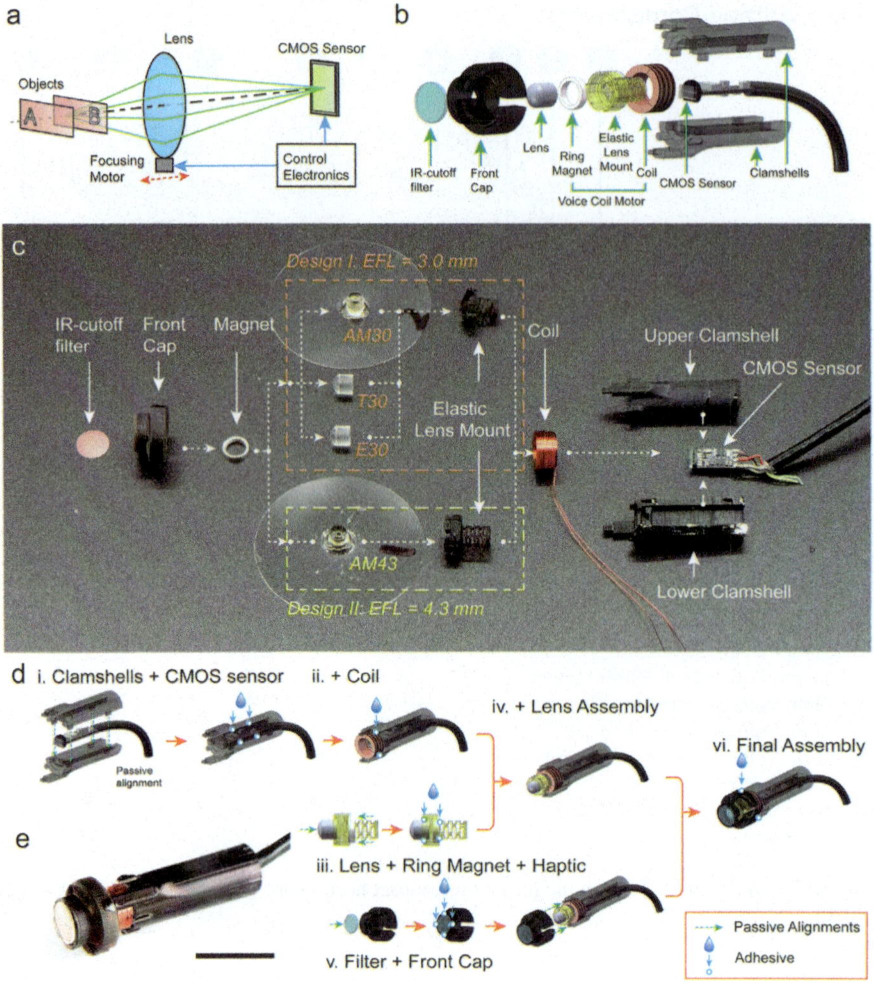

Fig. 4.53 A miniaturized microscope made in 50 mins at just US$4. Image is taken from Hai et al. (2023)

Examples of their AFIs also include a $2 hearing aid for children developed by the neurosciences center CNEURO (Reardon 2016).

The last entry is a case in point of the rigor, in general, necessary while infusing frugality into the creation and also the operation of products. Here is an AFI consisting of a set of low-cost sensors measuring pollutants such as NO_2, NO and SO_2 with particulate-size between 2.5 and 10 μm in the Nairobian environment in Kenya (Lewis and Edwards 2016; deSouza et al. 2017). A group of researchers delved into the efficacy of such low-cost sensors, which despite their sophistication are warranted to be applied in cases of high concentration of particulate and also for kindling citizen awareness about pollution-related dangers. Consequently, these AFIs have to complement the conventional more accurate versions until rigorous calibrations and other procedures are carried out in the foreseeable future to make them replace their conventional counterparts.

4.5.1 AFIs to Tackle COVID-19

The smartphone has turned out to be a versatile AFI, in the hands of able researchers, for tackling the ongoing pandemic. The resulting positive impact has been achieved in two ways. First, smartphone-enabled scientific tools were used for battling the virus, and second, smartphone-based apps are used by people in general for gathering epidemiological data about the pandemic. Both these methods are very cost-effective, which is very important in reaching out to the wider world in an affordable manner. Dr. Debojyoti Chakraborty of the *Institute of Genomics and Integrative Biology*, part of one of India's top scientific bodies the *Council of Scientific and Industrial Research* (CSIR), has used CRISPR gene editing technology to develop bands on strips indicating the COVID-19 RNA. These bands are later studied by a smartphone app called TOPSE that photographs these bands and uses *machine learning* to detect viral infection (Ravindran 2021). There are also smartphone-based tools like portable fluorescence microscope, which costing ~\$40 is much cheaper than conventional benchtop microscopes costing ~\$15,000. Assembled from off-the-shelf components coupled with a smartphone camera, this portable microscope is about to go under clinical trials for detecting norovirus and also COVID-19 (Ravindran 2021). Another smartphone-based diagnostic tool for COVID-19 has been developed by Dr. Melanie Ott of the *Gladstone Institute of Virology and Immunology*, San Francisco along with Professor Daniel Fletcher of the *University of California*, Berkeley (Ravindran 2021). They claim that their smartphone-based test is faster and also comparable to standard ones. Moreover, their tool can even forecast transmission risk and severity; both of which are crucial factors in the epidemiology of this disease. As for apps, crowdsourced versions, involving millions of users, have facilitated scientific decisions on the use of masking according to Professor John Brownstein of the *Harvard Medical School* in Massachusetts, USA (Ravindran 2021). Lastly, Professor Huimin Zhao's group at UIUC has developed a ~\$90 AFI called SPOT, for *Scalable and Portable Testing*, by *3D printing* many of its components (Xun et al. 2021). SPOT, shown in Fig. 4.54, is based on a new rapid assay technique operating on saliva and delivering diagnostic results in about 30 min. The AFIs alluded to in this subsection are listed in Table 4.3.

Table 4.3 *Advanced Frugal Innovations* fighting COVID-19

No	Advanced Frugal Innovation (AFI)	Manufacturer (Country)	Frugal features	Compact design	Price
1	Low-cost Detection (Ravindran 2021)	Paper-Based Detection of Viral RNA (*Institute of Genomics and Integrative Biology* Delhi, India)	Uses CRISPR to create bands on paper strip as proxy for viral RNA. Subsequently, an app images the bands and uses machine learning to detect infection	Yes	Low cost
2	Low-cost Fluorescence Microscopy (Ravindran 2021)	Portable Fluorescence Microscope (University of Arizona Tucson, USA)	Coupling of smartphone with off-the-shelf microscope components along with MATLAB is facilitating detection of norovirus and COVID-19 at low cost.	Yes	Just ~$40!
3	Low-cost Diagnostic (Ravindran 2021)	Smartphone-based Diagnostic (Gladstone Institute of Virology and Immunology, University of California Berkeley, USA)	Use of smartphone for detecting COVID-19 at low cost. It also detects transmission risk and disease severity.	Yes	Low cost
4	Epidemiological App (Ravindran 2021)	Smartphone App (Harvard Medical School, USA)	Use of apps for collecting data from society at large. Has helped reach decisions on masking!	Yes	Low cost
5	Rapid Diagnostic Test (Xun et al. 2021)	*3D Printed* Assay-Device (University of Illinois at Urbana Champaign USA)	A device with *3D Printed* parts can diagnose COVID from saliva using a new assay technique in about 30 min	Yes	Just ~$90

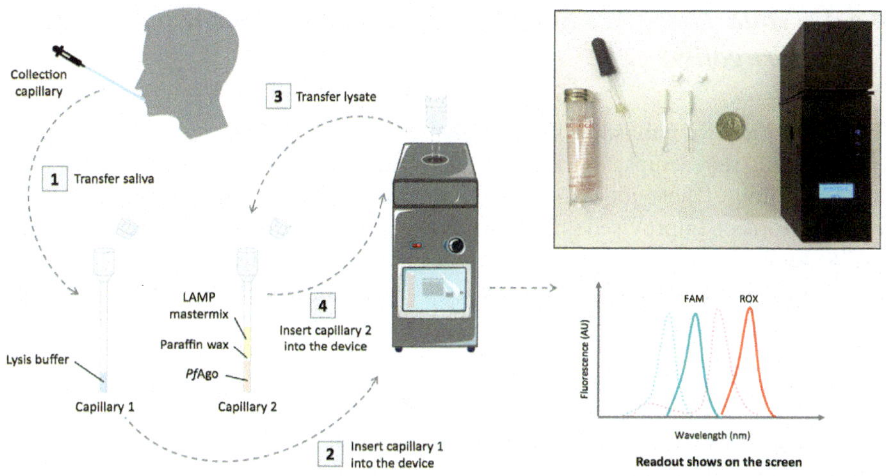

Fig. 4.54 Meet SPOT, a low-cost AFI giving COVID diagnostic results in about 30 min. Image is taken from Xun et al. (2021)

4.5.2 India's IT Sector!

The Indian *Information Technology* (IT) sector is a paramount example of an *Advanced Frugal Innovation* as a service. The entire boom in the IT sector started due to the cost-effective but superior quality IT services provided to companies world over. Three Indian IT giants that pioneered this AFI and stand out against global competition are *Infosys*, *Tata Consultancy Services* (TCS) and *Wipro*. Although the boom years of the 90 s were followed by decent progress, the pandemic has again buoyed the Indian IT business due to the need for remote and/or hybrid working conditions (Economist 2022).

4.5.3 Nature-Based Solutions

The heading of this subsection is a new acronym "NBS" that has found its way into a *Nature* editorial (Nature-Editorial 2017). *Nature-Based Solutions* according to Nature-Editorial (2017) refer to sustainable practices that utilize services of nature or the natural world. An example of NBS is leveraging wetlands for cleaning water. The concept of *advanced frugal innovations* gels well with NBS in that nature-based principles can be applied to create devices that partly/wholly use nature's services in their functionality. This is akin to biomimetics and nature-based principles could help engineer AFIs leveraging nature's services for affordably carrying out robust functionality in various sectors. Moreover, NBS itself will need to harness AFIs for its success in the new era of *climate change*.

4.5.4 Voilà There Are Scientific Journals Documenting Frugality

Two new journals published by *Elsevier* and *Ubiquity Press* now accept papers on *do it yourself* setups for carrying out various activities in all areas of science and technology. These journals, *Journal of Open Hardware* and *HardwareX*, peer-review efforts on improvised or DIY setups that contribute to advancement in science and technology. A primary criterion for acceptance is that these devices have to be affordable. The foreseeable future will see the coming of more such journals and plausibly some on *frugal engineering*!

4.6 Ingredients for Making a Successful AFI

The innovations in Table 4.1 were either developed as a new product, such as entries [1–12, 14, 16–18, 20–25], or were retrofitted from existing designs and/or the recycling of *end-of-life* components as is evident in the fabrication of the remaining AFIs including the Square Kilometer Array for radio telescopy [19] (Nordling 2012), improvised submarine Stinky [15] (Davis 2014), and the robotic hand [13] (Greenemeier 2012). Notwithstanding their fabrication, these innovations are AFIs with their telltale features: realized by the application of S&T; workers are cognizant of S&T principles through formal training and they are *frugal.*

The examples covered in Table 4.2 are AFIs designed and fabricated from scratch. The reader should have attempted to discern the features, given in the definition at the beginning of this chapter, in the AFIs of Tables 4.1 and 4.2. Readers have to think through these examples to grasp the potential of an AFI. Such understanding will aid readers in following the remaining chapters on the engineering of these products.

4.6.1 Developers of Advanced Frugal Innovations

All of the innovations, except [1, 2, 15, 16, 22] in Table 4.1, were realized by workers with a strong background in *science and technology* with relevant details to be found in the references attached to each table entry. Even entries [2, 16, 22] involve mentoring or development of the original AFIs for DIY practitioners by workers with advanced degrees. This is because these innovations have been realized by *research and development* (R&D) teams, in academia and/or industry, where the knowledge base of the workforce needed to develop cars, internal combustion engines, accelerators, missiles and rockets, to name some of the innovations listed in Table 4.1, is accrued typically through advanced degree programs beyond a bachelor's degree. Also, end users as reported by von Hippel (1976) were not the innovators for the AFIs listed in Table 4.1. This is because von Hippel's observation applies to frugal innovations of the grassroots type, which are typically makeshift arrangements arrived at by the ingenuity of end users, as innovators, at the base of the pyramid (Gupta 2012, 2013).

Again, the AFIs of Table 4.2 were developed majorly by researchers and workers with scientific knowledge obtained through advanced degree programs. Science can do wonders and this book has shown examples where frugality has in fact furthered the cause of science. In other words, focusing the lens of frugality on any product would most of the time lead to a new and better version of the product and in the process advance science itself.

4.6.2 Research and Development on Advanced Frugal Innovations

As evidenced by the frugal features listed in Tables 4.1 and 4.2, budgetary constraints have encouraged researchers to seek novel solutions that are also advancing the fields of science and engineering. The examples of the plasma collider; palm-sized DNA sequencer; lightning sensor; aircraft wing and TOLIMAN telescope, to name a few, point to new science and technology expanding the frontiers of knowledge. Consequently, the conceptualization and realization of AFIs can benefit immensely from research and development programs dedicated to their creation. This is because an effective R&D program will aid in the systematic application of advances in S&T to realize frugal innovations of the advanced type. Furthermore, such research activity could also aid in advancing S&T.

4.6.3 Design Philosophy for Advanced Frugal Innovations

The continued use of frugality to create AFIs warrants a philosophy rooted in science and technology that will aid in the methodical realization of "more-with-less", a term coined by Prabhu (2017). A methodical approach is needed to improve the safety and performance of designs done-with-less, i.e., realized with lesser resources. Such a philosophy is also required to avoid misuse of the frugality approach by not least, profit-seeking entities taking it up in a haphazard manner. Moreover, this philosophy has to apply to the creation of AFIs as a novelty from scratch and/or the assembly of components salvaged from end-of-life systems.

The three essential elements of a low-cost engineering philosophy are the following. First, the emerging area of *frugal engineering* is interdisciplinary which embraces various fields of S&T. This is because advances in S&T, in general, are used to advantage for lowering the costs of sophisticated products, which will facilitate inclusive growth. In this regard, the AFIs listed in Tables 4.1 and 4.2 attest to the use of progress in various areas of S&T for reducing costs. Therefore, research is needed to keep abreast of the latest developments in science and engineering, in general, and later to extract ideas from this large pool that are relevant to the creation of a given AFI. The combing of such a wide swathe of scientific fields is essential to hone in on relevant portions of traditional sciences and/or emerging fields. An instance of an emerging field in engineering is Biomimetics wherein nature-based frugal principles are being brought to bear on realizing products with judicious use of resources that perform effectively (Ball 2001).

Second, the leveraging of advances in S&T for creating AFIs would be incomplete without the rigor in subsequent design in view of the typical frugal amounts of resources involved. Therefore, a rigorous procedure should be established that minimizes the uncertainties involved in the design process. The minimization of uncertainties, related to aspects of design such as materials and manufacturing, would result in the best-case scenario of lowering the *factor of safety* (S) to $\sim$1.5. The S value of 1.5 is routinely employed in the aircraft industry (Shanley 1962; Norton 2006) along with rigorous design procedures to achieve robust aircraft with lower weight for effective flight. Consequently, the design of AFIs warrants researches in systematizing rigorous design methodology, accompanying a lower S, that will lower various uncertainties thereby minimizing resource consumption. This entails R&D for identifying and later improving various aspects of design so as to minimize their uncertainties for achieving rigor with suitable trade-offs that do not hinder the functionality of the product (Otto and Antonsson 1991). Some of the aspects alluded to include the following: accurate design and material models; near-net-shape and sustainable manufacturing processes; various failure events including those with a lesser probability; and extension of standardization of parts and other systems for inclusion of a wider range in shapes and sizes. Furthermore, such a rigorous design methodology should be applied iteratively to progressively iron out the various weak spots in a design that possibly escape notice in a single pass of the design process. Although this second element has been proposed from the perspective of engineering, the importance of rigor in scientific endeavors, not based on the safety factor approach, is not lost. Overall, rigor in all knowledge-based endeavors creating AFIs is imperative when avoiding excesses in the consumption of resources.

And third, a strong nexus should be maintained with entities actively involved in developing frugal innovations for the lower income segment of society, to study their ingenious incorporation of thrift in grassroots products (Gupta 2012, 2013) and, where possible, facilitate its extension to AFIs. In fact such studies on the lower income segment should be undertaken in developed and developing countries alike to avoid missing out on the next big idea. Although the content here can be subsumed under general research mentioned in the second element, its isolation as a third element is sought to highlight the significance of thrift in any frugal innovation, grassroots, or advanced type.

4.7 A Caveat on the Use of Frugality

The "frugalization" of a product design should be thoroughly vetted for its feasibility during conceptualization, especially in applications where the human life is at stake. Subsequently, the low-cost philosophy presented in this effort, with room for futuristic revisions, should be followed for the systematic realization of AFIs, whose performance is not sacrificed by their *frugal* nature. Apart from the involvement of the human angle, the careful creation of AFIs is warranted generally for their eventual quality and hence success.

Other factors can also play a role in the failure of a *frugal innovation*, an instance of which is the low-cost laptop developed with inclusive growth in mind. The laptop developed initially for children through an initiative and movement called the *One Laptop Per Child* (OLPC)—http://one.laptop.org/—was an amazing piece of brilliant hardware (see *problem 3*) (Robertson 2018). Other than lacking a stringent quality check that is imperative to *frugal* products, the failure of the OLPC was also rooted (in my opinion) in the lack of commensurate sales and political nous. The concept of a cheap laptop became so endearing that other entities including private ones also subsequently joined the bandwagon. These private sector entities sold their wares and profited, because of their honed commercial skills, in the wake of the OLPC movement. This example on the OLPC stresses the fact that the successful diffusion of AFIs in society at large needs careful navigation of the design and engineering, marketing, socioeconomic and political landscapes.

From an engineering point of view, caveats associated with frugality, therefore, necessitate the rigorous and systematic creation of AFIs since their quality of performance hinges on careful preparation due to constraints on all kinds of resources. In this regard, the chapters to follow describe a new engineering philosophy called *frugal engineering* involving *frugal design* and *frugal manufacturing* that are rooted in a modern version of the *classical factor of safety* known as the *factor of frugality*.

Problems on Identifying Innovations

1. The *National Aeronautics and Space Administration* (NASA) has teamed up with the company *AutoDesk* in designing a rover for the moons of Jupiter and Saturn. The rover will have to experience more challenging environments than those that went to Mars. Critical to this mission is a rover structure that is low in weight but strong in mechanical properties. Accordingly, the structure was conceptualized using *Autodesk's* Generative Design that uses *artificial intelligence* and fabricated plausibly with *additive manufacturing* to arrive at the rover shown in Fig. 4.55. This is an AFI which has not been listed in the tables of this chapter. Elaborate on the features of the rover that make it an AFI. **Hint: Literature survey is necessary to study and list these features!**

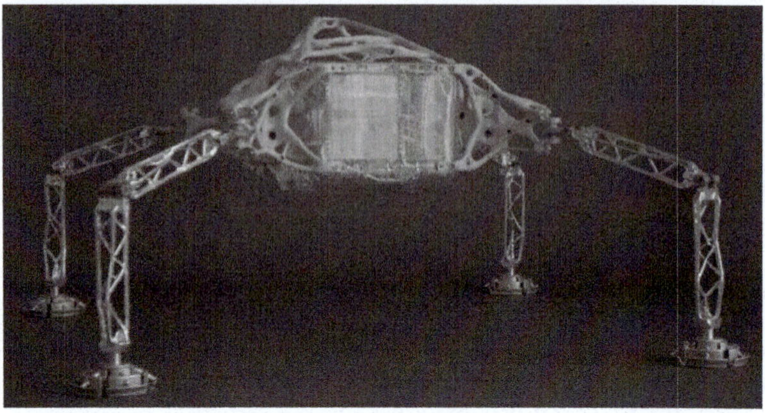

Fig. 4.55 A rover conceptualized through *Generative Design*. Image is taken from https://www.machinedesign.com/3d-printing-cad/article/21837312/nasa-uses-generative-design-on-nextgeneration-interplanetary-lander

Fig. 4.56 A unmanned robotic U.S. Navy boat making sophisticated affordable defense against enemies possible. Image is taken from Popular Mechanics (2022)

2. Farmer *Anand* has individually used three kinds of fertilizers A, B and C to determine their efficacy in warding off *weeds*. The samples of A, B and C, given by a friend, were applied haphazardly in some segragated patches of *Anand's* field so that no two fertilizer samples had any interaction effects. *Anand's* friend claims that these fertilizer samples are not only cost-effective, but also made by top-notch companies using the latest technology. Will the winning fertilizer sample be a GFI or an AFI?
3. The final year undergraduate students of engineering at the *Indian Institute of Technology Madras* have designed and fabricated payloads which are low-cost sensor-based experiments to be carried on CubeSats. Other than the rigorous curriculum in science, mathematics and engineering, these students have also taken advanced level courses related to designing such satellites and their payloads. Do their low-cost contraptions qualify as an AFI?
4. Professor Negroponte of the *Massachusetts Institute of Technlogy* (MIT) developed a low-cost, i.e. $100, laptop for the low-income and/or poorer segments of the world called XO-1 through the *One Laptop Per Child* (OLPC) movement (Robertson 2018). Professor Negroponte and his research group at the MIT Media Lab developed this laptop with a hand-cranking mechanism, in the initial version, for powering the laptop without electrical means. Robustness was also built along with other computer-related features for low-cost applications in rugged parts of the world. Would you consider the OLPC as an AFI?
5. Anurudh Ganesan, who was 17-year-old (when he made this contraption), has designed a human-driven cycle cart called *Vaxxwagon* that carries vaccines (Popular Mechanics 2015). A critical thing for vaccines is to keep them refrigerated through high-cost refrigerated trucks for example. Young Anurudh's cycle cart uses human pedaling to power a small water cooler mechanism through mechanical action. Designed and fabricated by the knowledgable Anurudh, the *Vaxxwagon* can carry vaccines thereby giving a low-cost solution to an important issue, especially in the context of pandemics! Is the under-$100 *Vaxxwagon* an AFI?
6. AFIs have also appeared in defense applications and an example on a missile appeared in Table 4.1. Yet another robotic-AFI recently announced by the U.S. Navy is the *medium unmanned surface vehicle* (MUSV) shown in Fig. 4.56. About 12 MUSVs can replace a manned $1.8 billion *Burke*-class destroyer and patrol 12 times as much water. Moreover, MUSVs also save the additional cost of employing 400 crew members for operating a *Burke*-class destroyer. In contrast, each MUSV costs $150 million which carry out their tasks effectively. Explain why a MUSV qualifies as an AFI. **Hint: The reference for this question is this article in** Popular Mechanics (2022).
7. Undergraduate students at one of the finest engineering schools in India have developed a model aircraft out of *balsa* wood. They have *Jugaaded* a lot in attempting to keep costs low and also due to their limited access to sophisticated resources. This venture was aided significantly by several relevant courses taken by these students as part of their undergraduate curriculum in engineering. These courses, related to the design of aircraft; aircraft dynamics; propulsion and other

advanced level undergraduate courses, significantly improved the understanding of the working and design of aircraft. So is the resulting model aircraft a flying *Jugaad*?

8. Scientists at a fertilizer company were pursuing a new chemical cocktail for clearing up invasive weeds in agricultural fields. They wanted to make an affordable cocktail effective in attacking weeds in both the rich and poor regions of the world. However, try as they might they could not get the chemical elements required and their mixing-ratios. In frustration, they resorted to just the brute force approach of trying various chemical combinations until they got the winning formula called KillWeed. Is KillWeed an AFI?
9. Which of the AFIs listed in this chapter interest you the most and why? You could list more than one. Also, describe how frugality has made a deep impact on this product(s) when compared to its conventional brethren.
10. List at least one AFI from your field of interest that is not listed here. You could convey a brief write-up to the author at your convenience, which could be included in future editions of this book.

References

Airforce-Technology (2021) Continuous composites completes AFRL's WiSDM contract for LCAA wing

Arenaev (2023) Tesla explains its het pump technology in new promo video. https://www.arenaev.com/tesla_explains_its_heat_pump_technology_in_new_promo_video-news-1326.php

Arnold C (2019) Medicines on demand. Nature 575:274–277

arsTECHICA (Aug 2020) Android is now the world's largest earthquake detection network. https://arstechnica.com/gadgets/2020/08/android-is-now-the-worlds-largest-earthquake-detection-network/

Aurolab (n.d.) https://aravind.org/aurolab/

Austen K (2015) Pollution patrol. Nature 517:136–138

AWAKE (2016) Awake explores the use of plasma to accelerate particles to high energies over short distances. https://home.cern/science/accelerators/awake

Awtar S, Truant T, Nielsen J, Abani R (2010) Flexdex TM: a minimally invasivesurgical tool with enhanced dexterity and intuitive control. ASME J Med Devices 4:035003

Ball P (2001) Life's lessons in design. Nature 409:413–416

Bao J, Bawendi M (2015) A colloidal quantum dot spectrometer. Nature 523:67–70

BBC-News (2018) Low-cost 'four-hour' bamboo house wins top prize

BikeBandit (May 2018) Drum chargers: cheap bolt-on forced induction that actually works? https://www.bikebandit.com/blog/drum-chargers-cheap-bolt-on-forced-induction-that-actually-works

Byagathvalli G, Pomerantz A, Sinha S, Standeven J, Bhamla M (2019) A 3d-printed hand-powered centrifuge for molecular biology. PLoS Biol 17(5):e3000251

Cressey D (2017) Toolbox: age of the Arduino. Nature 544:125–126

Cruccolini V, Discepoli G, Zembi J, Battistoni M, Mariani F, Grimaldi C (2018) Experimental assessment of a pressure wave charger for motorcycle engines. Energy Procedia 148:1254–1261

Cybulski J, Clements J, Prakash M (2014) Foldscope: origami-based paper microscope. PLoS One 9(6):e98781

Davis J (2014) Spare parts: four undocumented teenagers, one ugly robot, and the battle for the American dream. Farrar, Straus & Giroux, New York, NY, USA

deSouza P, Nthusi V, Klopp JM, Shaw BE, Ho WO, Saffell J, Jones R, Ratti C (2017) A Nairobi experiment in using low cost air quality monitors. Clean Air J 27(2):12–42

Dey R, Patnaik S, Steiner S, Sivarasu S (2016) Low-cost three-dimensional printed surgical drill-guiding device for MPFL reconstruction (Pat-Rig). ASME J Med Devices 10(2):020914

Downer M, Zgadzaj R (2014) Surf's up at SLAC. Nature 515:40–41

Economist (17 Apr 2010) First break all the rules: the charms of frugal innovation

Economist (2012) Rockets galore: cheap smart weapons. The Economist

Economist (2014) MOMs are from Mars

Economist (2015) Technology quarterly: difference engine—the little engine that could

Economist (2020) Schumpeter: which Japanese mogul will leave the biggest legacy?

Economist (2021) A Japanese forestry firm wants to put wooden satellites into orbit

Economist (2022) A half-a-trillion-dollar bet on revolutionising white-collar work

Electrek (2020) Tesla model y has a heat pump for consistent range in cold climates. https://electrek.co/2020/03/13/tesla-model-y-has-a-heat-pump-for-consistent-range-in-cold-climates/

Esteva A, Kuprel B, Novoa R, Ko J, Swetter S, Blau H, Thrun S (2017) Dermatologist-level classification of skin cancer with deep neural networks. Nature 542:115–118

Extance A (2015) Laser weapons get real. Nature 521:408–410

Garcia M, Edmiston C, York T, Marinov R, Mondal S, Zhu N, Sudlow GP, Akers WJ, Margenthaler J, Achilefu S, Liang R, Zayed MA, Pepino MY, Gruev V (2018) Bio-inspired imager improves sensitivity in near-infrared fluorescence image-guided surgery.Optica 5(4):413–422

Gibney E (2015) News: CERN to test revolutionary mini-accelerator. Nature 526:173–174

Gibney E (2016) News in focus: open-hardware' pioneers push for low-cost lab kit. Nature 531:147–148

Gibney E (2021) News feature: the quest to understand how thunderstorms create gamma rays. Nature 590:378–381

Gilbert N (2016) Frugal-farming. Nature 533:308–310

GreenCarCongress (2019) Brunel to use camcon single cylinder IVT in researching future power-train concepts. Green Car Congress

Greenemeier L (2012) Four-fingered robot can replace flashlight batteries. Nature 488:11238

Grimminger F, Meduri A, Khadiv M, Viereck J, Wüthrich M, Naveau M, Berenz V, Heim S, Widmaier F, Flayols T, Fiene J, Badri-Spröwitz A, Righetti L (2020) An open torque-controlled modular robot architecture for legged locomotion research. IEEE Robot Autom Lett 5(2):3650–3657

Gupta A (2012) Innovations for the poor by the poor. Int J Technol Learn Innov Dev 5(1/2):28–39

Gupta A (2013) Tapping the entrepreneurial potential of grassroots innovations. Stanf Soc Innov Rev 11(3):18–20

Hai R, Shao G, Ware H, Jones E, Sun C (2023) 3d printing a low-cost miniature accommodating optical microscope. Adv Mater, 2208365

Hayden E (2015) Pint-sized DNA sequencer impresses first users. Nature 521:15–16

Heiden A, Preninger D, Lehner L, Baumgartner M, Drack M, Woritzka E, Schiller D, Gerstmayr R, Hartmann F, Kaltenbrunner M (2022) 3d-printing of resilient biogels for omnidirectional and exteroceptive soft actuators. Sci Robot 7:eabk2119

IKEA (Dec 2022) Ikea explores multifunctional furniture for small spaces. https://www.ikea.com/global/en/stories/design/ikea-explores-multifunctional-furniture-to-push-limits-of-small-space-221206/

Immelt J, Govindarajan V, Trimble C (2009) How GE is disrupting itself. Harv Bus Rev 87(10):56–65

Irving M (Aug 2020) Prototype gravity-based energy storage system begins construction. https://newatlas.com/energy/gravitricity-gravity-renewable-energy-storage-system/

Janoschka T, Martin N, Martin U, Friebe C, Morgenstern S, Hiller H, Hager MD, Schubert US (2015) An aqueous, polymer-based redox-flow battery using non-corrosive, safe, and low-cost materials. Nature 527:78–93
Jeff T (2014) Islands of light. Nature 507:154–156
Jiang Y, Korpas L, Raney J (2019) Bifurcation-based embodied logic and autonomous actuation. Nat Commun 10(128)
Jones N (2014) Mini satellites prove their scientific power. Nature 508:300–301
Jones N (2016) First flight for tiny satellites. Nature 534:15–16
Joyce A (2012a) The TB test you can do at home. Nature, 11327
Joyce A (2012b) The TB test you can do at home. Nature, 11327
Kay R, Katrycz C, Nitièma K, Jakubiec JA, Hatton BD (2022) Decapod-inspired pigment modulation for active building facades. Nat Commun 13(4120)
Kumar S (2014) News: Indian mars craft prepares for orbit. Nature 513:291
Lee D (Jun 2019) Ikea is introducing robotic furniture for people who live in small spaces. https://www.theverge.com/2019/6/4/18652178/ikea-rognan-robot-murphy-bed-furniture-ori-living-democratic-design-days
Leeming J (2022) Where i work: Cliff falls: using a smartphone to track erosion. Nature 604:794
Lesté-Lasserre C (2020) This $1 hearing aid could treat millions with hearing loss. Science
Lewis A, Edwards P (2016) Comment: validate personal air-pollution sensors. Nature 535:29–31
Liu C, Kong D, Hsu P, Yuan H, Lee H, Liu Y, Wang H, Wang S, Yan K, Lin D, Maraccini P, Parker K, Boehm A, Cui Y (2016) Rapid water disinfection using vertically aligned mos2 nanofilms and visible light. Nature 11(12):1098–1104
Liu H, Huang Y, Jiang H (2016) Artificial eye for scotopic vision with bioinspired all-optical photosensitivity enhancer. Proc Natl Acad Sci 113(15):3982–3985
Long KD, Woodburn EV, Le HM, Shah UK, Lumetta SS, Cunningham BT (2017) Multimode smartphone biosensing: the transmission, reflection, and intensity spectral (tri)-analyzer. R Soc Chem: Lab Chip 17:3246–3257
Maxmen A (2016) Big pharma's cost-cutting challenger. Nature 536:388–390
Medeiros MD, Chanci D, Martinez RV (2020) Moisture-insensitive, self-powered paper-based flexible electronics. Nano Energy, 105301
Middlemiss RP, Samarelli A, Paul DJ, Hough J, Rowan S, Hammond GD (2016) Measurement of the earth tides with a mems gravimeter. Nature 531:614–617
Miesler T, Wimschneider C, Brem A, Meinel L (2020) Frugal innovation for point-of-care diagnostics controlling outbreaks and epidemics. ACS Biomater Sci Eng 6(5):2709–2725
Miller O (2021) Bubbling to the surface: WVU engineers develop new geothermal energy technology
Nature (2013) Research highlights: mobile worm microscope. Nature 495:285
Nature (2014) Nature's10: robot-maker. Nature 516:314
Nature (2015) Research highlights: cheap MRI uses small magnets. Nature 526:479
Nature (2019) Research highlights: mini-factory churns out drug cheaply and at a furious pace. Nature 577:148
Nature (2021) A watchful eye on the deep-sea environment
Nature (2022) Research highlights: flower power: 'sunflower' system churns out energy. Nature 605:399
Nature-Editorial (2017) Editorial: natural language. Nature 541:133–134
Nature-Editorial (2018) Editorial: smelting point. Nature 557:280
Nordling L (2012) News: recycled dishes form telescope network. Nature 488:488571
Nordling L (2013) News feature: southern star. Nature 502:608–610
Norris, G. (August 2020) , 'Rolls-royce reveals attritable engine core demo.', Aviation Week Network
Norton R (2006) Machine design: an integrated approach. Pearson Prentice Hall, Englewood Cliffs, New Jersey, USA
Otto K, Antonsson E (1991) Trade-off strategies in engineering design. Res Eng Des 3(2):87–104

Padma T (2017) News: India gears up for second moon mission. Nature 550:440

Pardee K, Green AA, Takahashi MK, Braff D, Lambert G, Lee JW, Ferrante T, Ma D, Donghia N, Fan M, Daringer NM, Bosch I, Dudley DM, O'Connor DH, Gehrke L, Collins JJ (2016) Rapid, low-cost detection of zika virus using programmable biomolecular components. Cell 165:1255–1266

Pavlovich M (2014) Antimicrobial applications of ambient–air plasmas, PhD dissertation, University of California at Berkeley

Popkin G (2018) Tabletop physics pushed to the edge. Nature 553:142–144

Popular Mechanics (2015) 15 young innovators who are already making a difference: introducing the next generation of brilliant minds. Popular Mechanics

Popular Mechanics (2016) How cataract surgery went from $500 to under 2$. Popular Mechanics

Popular Mechanics (2022) The navy's latest Robo-Ship will test out a 'ghost fleet' concept. Popular Mechanics

Prabhu J (2017) Frugal innovation: doing more with less for more. Philos Trans R Soc A: Math Phys Eng Sci A375:20160372

Prasad A, Hasan SMA, Grouchy S, Gartia MR (2019) DNA microarray analysis using a smartphone to detect the BRCA-1 gene. R Soc Chem: Anal 144:197–205

Quick J, Loman N, Duraffour S et al (2016) Real-time, portable genome sequencing for Ebola surveillance. Nature 530:228–232

Quinn M, Bubi TC, Pierce MC, Kayembe MK, Ramogola-Masire D, Richards-Kortum R (2012) High-resolution microendoscopy for the detection of cervical neoplasia in low-resource settings. PLOS One 7(9):e44924

Rafalskyi D, Martínez J, Habl L, Rossi EZ, Proynov P, Boré A, Baret T, Poyet A, Lafleur T, Dudin S, Aanesland A (2021) In-orbit demonstration of an iodine electric propulsion system. Nature 599:411–415

Rao B (2017) Advances in science and technology through frugality. IEEE Eng Manag Rev 45(1):32–38

Rao B (2017) Leading edge: advanced frugal innovations. IEEE Technol Soc Mag 36(4):53–54

Rao B, Bhinge R (2013) Development of a pedal powered water filtration system (filed for Indian patent: 3046/che/2013)

Ravindran S (2020) Work/technology & tools: how DIY technologies are democratizing science. Nature 587:509–511

Ravindran S (2021) Technology feature: smartphone science apps test and track infectious diseases. Nature 593:302–303

Reardon S (2016) Cuban science at a crossroads. Nature 537:600–603

Research highlights: paper-based gene tools (2014) Nature 514:539

Robertson A (Apr 2018) Olpc's 100-dollar laptop was going to change the world - then all went wrong. https://www.theverge.com/2018/4/16/17233946/olpcs-100-laptop-education-where-is-it-now

Sapra NV, Yang KY, Vercruysse D, Leedle KJ, Black DS, England RJ, Su L, Trivedi R, Miao Y, Solgaard O, Byer RL, Vučković J (2020) On-chip integrated laser-driven particle accelerator. Science 367(6473):79–83

Scientific American (2015) Spacex faces the hard truth about so3 landings—they're tough to do. Scientific American

Scientific American (2016) Maker movement turns scientists into tinkerers. Scientific American

Shah A (Apr 2018) Clean clothes green. https://www.asme.org/topics-resources/content/clean-clothes-green

Shanley F (1962) Historical note on the 1.5 factor of safety for aircraft structures. J Aerosp Sci 29(2):243–244

Song G, Chu K, Kim S, Crose M, Cox B, Jelly ET, Ulrich JN, Wax A (2019) First clinical application of low-cost oct. Transl Vis Sci Technol

Space (Jun 2019) Tiny chipsats, big success: cracker-size probes phone home from orbit. https://www.space.com/tiny-chipsats-ace-demonstration-mission.html

Stanford (n.d.) Onebreath: a more affordable ventilator: an interview with matt callaghan of onebreath. https://biodesign.stanford.edu/our-impact/technologies/onebreath.html
Stone R, Kelly D, Geddes J, Jenkinson S (2017) Intelligent valve actuation—a radical new electromagnetic poppet valve arrangement. Aachen Colloquium
Sun F, Ganguli A, Nguyen J, Brisbin R, Shanmugam K, Hirschberg D, Wheeler MB, Bashir R, Nash D, Cunningham B (2020) Smartphone-based multiplex 30-minute nucleic acid test of live virus from nasal swab extract. R Soc Chem: Lab Chip 20:1621–1627
Tarancón A (2019) Powering the IoT revolution with heat. Nature 2(7):270–271
techcrunch (2021) Eli evs begins production, European rollout of Eli zero, a compact two-seater. https://techcrunch.com/2021/09/20/eli-evs-begins-production-european-rollout-of-eli-zero-a-compact-two-seater/
Technology-Review (2011) The next generation of technology: 35 innovators under 35
This device could detect dozens of cancers with a single blood test (2014)
Tollefson J (2013) Lightning network tested out in guinea. Nature 502:604–605
Tollefson J (2017) News: mobile-phone signals bolster street-level rain forecasts. Nature 544:146–147
Tuthill P, Bendek E, Guyon O, Horton A, Jeffries B, Jovanovic N, Klupar P, Larkin K, Norris B, Pope B, Shao M (2018) The TOLIMAN space telescope, vol 10701. Society of Photo-Optical Instrumentation Engineers (SPIE), Austin, Texas, USA
UnivofSydney (2021) Worlds next door: looking for habitable planets around alpha centauri
von Hippel E (1976) The dominant role of users in the scientific instrument innovation process. Res Policy 5(3):212–239
Vorraro G, Turner M, Turner JWG (2019) Testing of a modern Wankel rotary engine—part i: experimental plan, development of the software tools and measurement systems. SAE Technical Paper (2019-01-0075), pp 1–16
Vos BE, Blesa EB, Betz T (2021) Designing a high-resolution, lego-based microscope for an educational setting. Biophys 2(3):29–40
Weiss C (Sep 2016) British roadster prototype addresses rotary engine shortcomings. https://newatlas.com/aie-rotary-roadster/45514/
Wild S (2017) News in focus: Ghana telescope ready for action. Nature 545:144
Xun G, Lane S, Petrov V, Pepa B, Zhao H (2021) A rapid, accurate, scalable, and portable testing system for covid-19 diagnosis. Nat Commun 12(2905)
Yeh Y-T, Gulino K, Zhang Y, Sabestien A, Chou T-W, Zhou B, Lin Z, Albert I, Lu H, Swaminathan V, Ghedin E, Terrones M (2020) A rapid and label-free platform for virus capture and identification from clinical samples. Proc Natl Acad Sci (PNAS) 117(2):895–901
Zhakypov Z, Mori K, Hosoda K, Paik J (2019) Designing minimal and scalable insect-inspired multi-locomotion millirobots. Nature 571:381–386
Zhao L, Bhatia B, Zhang L, Strobach E, Leroy A, Yadav MK, Yang S, Cooper TA, Weinstein LA, Modi A, Kedare SB, Chen G, Wang EN (2020) A passive high-temperature high-pressure solar steam generator for medical sterilization. Joule 4:1–13

Chapter 5
Engineering of Frugal Products

5.1 Introduction

The introduction to *Advanced Frugal Innovations* in the previous chapter should have convinced the astute reader that engineering is very important for infusing affordable quality performance in real-time applications. In fact, AFIs such as the compact *plasma* accelerator described in the previous chapter are impossible to realize even with just traditional engineering. The use of cutting-edge research that applies engineering principles in new directions has made this portable accelerator possible. Therefore, the principles of engineering and the advances we continue to make in both engineering and other fields are significant to every stage of product development involved in creating successful AFIs!

5.2 Need for Engineering

The makeshift feature of *grassroots frugal innovations* limits the potential of *frugal innovations* in general. This is because the absence of an understanding of the physics underlying the working of any product can lead their creators astray and also plausibly give lower quality performance to the consumer. A makeshift product's success should be attributed to the ingenuity of its creators, who though not trained in any scientific or other disciplines are still able to put together a product under conditions of natural scarcity. Therefore, it is worth pondering over what such brilliant creators can accomplish when trained in *science and technology* (S&T) or at least have access to experts in these areas.

Despite the success of many grassroots innovations, generally *frugal innovations* have been receiving flak due to their initial foray into makeshift products. Although, *grassroots frugal innovations* continue to thrive, a majority of FIs now coming into the market are of the *advanced* type that are systematically created from an artificial constraint on resources by focusing on minimizing costs. Hence, significant engi-

© The Author(s), under exclusive license to Springer Nature Singapore Pte Ltd. 2024
B. C. Rao, *Frugal Engineering*, Design Science and Innovation,
https://doi.org/10.1007/978-981-99-9700-8_5

neering is required to artificially constrain the consumption of material resources in an endeavor to realize products that are affordable while not crimping on functionality. From here onwards, the foundation of engineering itself has to be rooted in frugality for bringing about significant *sustainable development* on this planet. Moreover, mass production of AFIs in, ideally, all sectors of an economy can only be made possible by educating the existing (older) and newer workforce about the tenets of *frugal engineering*. A makeshift program at the grassroots level can no longer address the mass cravings of a world stepping up to sustainably adapt to *climate change* and other crises.

5.2.1 Secondary Activities for Effecting Frugality

The phenomenon of *frugal innovation* can be seen as a reduction in complexity due to scarce resources in the following: materials and products, innovation processes and business models. Other than the engineering of products (including materials), which is the thrust of this book, the other activities of innovation processes, business models, entrepreneurship and management are also important from the perspective of frugality. This book maintains that engineering from scratch during product development is a primary activity to achieving frugality and others are secondary that will supplement the main support of engineering.

Frugal innovations can be effectively implemented through their engineering and the surrounding policies, business environment, politics, social structure and other intangibles will help in creating a conducive environment. Accordingly, Leliveld and Knorringa (2018) have characterized *frugal innovation* as a concept that not only adapts and transforms products, but also influences the social, institutional, organizational and political systems. Therefore, their effort correctly concurs that meeting the challenges of *frugal innovations*, AFIs in particular, needs a multidisciplinary approach. Busch (2021) has also mentioned the necessity of suitable policies for supporting the multiple actors "under the radar" who nonetheless make *frugal innovations* possible. They have also reported that such transformations are inevitable due to the meeting of two worlds, i.e.,*multinational companies* (MNCs) and local *frugal* innovators, which are generally in favor of *frugal innovations*. However, Rosca et al. (2018) have also cautioned against MNCs exploiting local *grassroots frugal innovators*. Bhatti (2012) has also mentioned such intangible transformations being brought about by the FI phenomenon. Leliveld and Knorringa (2018) also highlight that companies in emerging markets like China, India and Brazil are also trying out new business models and distribution systems to extend their reach into lower and middle-income segments for buttressing diffusion of their *frugal* products. They (Leliveld and Knorringa 2018) have further suggested the consideration of poverty for better design and diffusion of at least *grassroots frugal innovations* which has also been recommended in the initial chapters of this book. Rosca et al. (2018) raise concerns about complex interactions between local, national and global stakeholders of FIs negatively influencing sustainability. In particular, they discuss possible adverse impacts of environmental regulations being violated by commercial interests in FIs. However, AFIs are engineered from scratch to be sustainable and as such have environmental credentials built through the rigor of design and fabrication that should not be overlooked by the entities involved. Here again governments have to set strict regulations and lay out policies to focus companies' efforts on sustainability.

The principles of this book when applied diligently will aid in understanding and even achieving a successful tradeoff among these various intangible factors in making the phenomenon of *frugal innovations* a success. Although there is significant room for improving tradeoffs between various stakeholders, the engineering of *frugal innovations* against *climate change* and other crises is a serious concept that the world cannot afford to ignore. The importance of various stakeholders and their socioeconomic aspects notwithstanding, FIs have to rise to our current grave situation by incorporating cutting-edge science and technology for engineering products that affordably help us adapt to a changing world with quality products. Even richer nations cannot afford to squander their resources on making complicated and/or bulky products because the resources involved could be used to make more of quality but streamlined products that cater to the rich, middle-income and the lower-income segments of the global society. Radjou and Prabhu (2015) clearly bring out the importance of frugality in the making of various products by showcasing case studies of famous *multinational* (MNC) companies. Both (Rao 2013) and (Radjou and Prabhu 2015) unequivocally state the disruptive nature of *frugal innovations* which can unleash revolutionary products in all sectors including automotive, aerospace, healthcare, defense, and others. However, (Radjou and Prabhu 2015) limit the role of *research and development* and make the consumer central to their strategy for using frugality in engineering. In contrast, this book puts emphasis on R&D while serving both business and consumer interests. R&D is just indispensable to make a engineering case for going *frugal*!

5.3 Frugal Engineering

Frugal Engineering (FE) is engineering for the creation of products with these features:

1. Low cost
2. Lower material consumption in product realization
3. Excellent functionality.

These three features of FE have to be reflected in every aspect of engineering. In other words, the constituents of engineering such as and, not limited to, design, manufacturing and materials should be rooted in these tenets of *frugal engineering* for any engineering activity targeted toward mitigating and also adapting to *climate change* and other crises. The overarching definition of FE when implemented for the individual pillars of engineering result in newer approaches to design, manufacturing, etc. This is borne out by later chapters in this book that describe *frugal* ways of designing and manufacturing. Although, the author is working on frugalizing the remaining aspects of engineering, this is a momentous task that needs participation from many in the research community who are hungry for challenging problems!

Overall, accounting for these tenets of FE can result in new engineering tools/platforms through which products in several disparate sectors could be built from scratch. In fact, even existing products can be retrofitted to account for the principles of FE thereby successfully transforming them into the *frugal* type. The use of such tools would make *frugal innovations* come out of the "makeshift" or *Jugaad* closet and make them advanced for widespread applications in a new era.

Even for people without a frugal mindset, it is noteworthy that the world is heading toward using S&T for mitigating and also adapting to the ongoing planetary crises. In particular, guidelines from bodies such as IPCC are turning into mandates by governments around the world for doing business and other endeavors. In light of such mandates, adoption of *frugal engineering* can lead to many advantages for early movers into the *frugal* way of doing things. Actually many companies are in the *frugal*-bandwagon due to their adoption of some or all of the above three tenets but are not aware of *frugal engineering* whose principles can be helpful.

5.3.1 Birth of a New Factor

The tenets of FE are easily adapted into existing engineering principles through the creation of a new factor, the *factor of frugality*. The traditional *factor of safety* has ruled over engineering, since its formal inception in the 17th century and, continues to do so. The *safety factor* affects every aspect of product development and hence is also majorly responsible for the popularity of conservative or safe designs. Conservative designs have a higher *factor of safety*, ~2 to 5, that generally translates into a padded product with extra material compensating for the various uncertainties stemming in engineering activities.

Against the current backdrop, a new version of the *factor of safety* is required for this new era of *climate change* that intrinsically accounts for material-savings. In other words, a new form of the *factor of safety* for this era of *climate change* that intrinsically minimizes material consumption. Accordingly, this book presents the *factor of frugality* that subsumes the traditional *safety factor* and whose adoption in engineering would make it compulsory to aim for utmost material-savings in every stage of product development. More on the *factor of frugality* in the next chapter on *frugal design*, which describes the first-generation model for this new factor along with details on its computations and engineering-features that influence its selection.

The subsequent chapters describe the pillars of *frugal engineering* in detail. The best way to leverage the potential of *frugal engineering* is to use *frugal* versions of its individual pillars. In other words, *frugal* versions of design, manufacturing, quality control, etc. should be utilized in this sequence to create a part/product. This results in a product development process that has not left any stone unturned in going *frugal*.

However, individual pillars such as *frugal manufacturing* could be used as a stand alone approach in traditional engineering to improve the efficacy of manufacturing.

5.4 Need for Research

The need for research would be felt acutely when traditional engineering principles are applied to existing problems or new areas for infusing frugality. This is because viewing a problem from the lens of frugality leads to challenges when aiming to achieving the trio of targets: low-cost; low-resource-consumption; and robust functionality. Therefore, a majority of *frugal* applications would need a advanced workforce, plausibly with doctorates and masters degrees, who are creative and inventive to solve these problems with ingenuity using *research & development* (R&D). Even existing problems when undertaken for a frugal transformation would need R&D activities since a *frugal* version of an existing problem is a new one with its own set of challenges.

As seen in the chapter on *advanced frugal innovations*, cutting-edge research is observed in a majority of the AFIs listed there. In general, the tenets of FE have become appealing over the past decade due to the exponential progress in S&T involving AI, additive manufacturing, computers, mobilephony and increased-and-inclusive access to the *Internet*. Such large-scale progress in S&T have made it relatively easier to tackle the challenge of frugalizing all products against the specter of *climate change* and other crises. It would not be an exaggeration to state that human progress has created *climate change* and such progress, specially in S&T, can only alleviate our suffering and help us adapt to this new age. And, frugalization through FE is the S&T-tool that will make this change possible !

Frugal engineering can be implemented only through complete infusion of its tenets in all of its pillars. Accordingly the next chapter presents a frugal version of the first stage in product development or engineering of a product called design. Subsequent chapters will deal with frugaization of manufacturing and also quality control to achieve near completion of frugalizing engineering. It should be noted that the concepts underlying these chapters are new and present a first-generation version of their respective models or equations. The global research community should undertake the challenges of revising or developing next generation models and also frugalize other pillars of engineering not undertaken in this book.

5.5 Need For a Frugal Mindset

Even with the availability of all scientific tools and capability to carry out R&D, a designer/engineer needs to develop a *frugal* mindset for successful realization of the product. Although, the years of grooming in traditional design methodologies gives the much needed rich experience, such experience can also become an impediment in *frugal engineering*. Therefore, the designer/engineer has to channelise his/her classical design knowledge for arriving at *frugal* products that work well during service. In fact, the *factor of frugality*, to be introduced in the next chapter, is firmly rooted in classical principles for designing *frugal* products in the first place.

Some have innate frugality related traits that will immensely aid in developing these products in various sectors. Others can develop a sense of frugality by repeated application of the *factor of frugality* approach that will eventually hone their skills in this regard. And all the while it should be borne in mind that we are aiming for frugality to arrive at top-quality products that do not reek of being cheap or a *Jugaad*.

5.6 Some Caveats on Frugal Engineering

The reader might have observed that the principles of *frugal engineering* as propounded in this book are applicable to any product in any sector. While this is true theoretically, a few caveats are in order while applying the principles of this book. As with any engineering methodology, a practionerer gets better with ever increasing use of necessary principles. This is the case with the methods of FE and the reader is entreated to thoroughly understand these principles as well as the application at hand before plunging head on into it. Therefore, challenges to new approaches like FE can be surmounted by detailed understanding of problem; detailed understanding of tools of FE; and practice!.

Although success is always welcomed with much fanfare, it is important that even failures are not ridiculed and are thoroughly analyzed to arrive at the weak spots in applying FE principles. This is all the more important in FE where severe constraints on resources and rigor make it incumbent to thoroughly understand an application from the prism of frugality.

References

Bhatti, Y. (2012) What is frugal, What is innovation? Towards a theory of frugal innovation. Said Business School, Working Paper, Oxford, UK

Busch H (2021) Frugal innovation in energy transitions: insights from solar energy cases in brazil. Cambridge J Reg Econ Soc 14(2):321–341

Leliveld A, Knorringa P (2018) Frugal innovation and development research. Eur J Dev Res 30:1–16

Radjou N, Prabhu J (2015) Frugal innovation: how to do more with less. Profile Books Limited, United Kingdom

Rao B (2013) How disruptive is frugal? Technol Soc 35:65–73

Rosca E, Reedy J, Bendul J (2018) Does frugal innovation enable sustainable development? A systematic literature review. Eur J Dev Res 30:136–157

Chapter 6
Frugal Design Approach

6.1 Introduction

The engineering of *frugal* products depends significantly on their design. This is because, the design stage provides many opportunities to correct flaws in the product at this initial conceptual level. Since frugality depends on strict restraints on resource consumption, the design process affords opportunities to iron out as many "heavy-spots" as possible to facilitate subsequent engineering of the *frugal* product. Hence the design process is crucial for *frugal* products and this chapter outlines a new design approach for infusing rigor in the engineering of such products. The design principles outlined in this chapter relate to products that have been developed to date through traditional engineering and hence are applicable to any exercise in new product development or frugalization of existing devices. Therefore, the approach suggested in this chapter will be applicable to the *advanced frugal innovations* covered in Chap. 4. The astute reader might be wondering about the plausibility of this statement because Chap. 4 has covered a wide range of AFIs including *point of care devices* and others.

It should be noted that the approach prescribed in this chapter can be applied to portions or the whole AFI. Accordingly, the methodology prescribed here will help arrive at streamlined versions of portions or a complete product. Even for cases not involving direct design, the *material saved* parameters (to be described below) and the concept underlying the approach of this chapter can lead to significant savings in materials and costs. This book is based on the premise that design will majorly influence directly or indirectly *frugal engineering* activities of any sector.

6.2 Background

Nature, as a minimalist, has created most of the flora and fauna on earth with an eye on economizing the consumption of available raw materials (Ball 2009). In contrast, since the dawn of the industrial revolution, designers have been typically

© The Author(s), under exclusive license to Springer Nature Singapore Pte Ltd. 2024

B. C. Rao, *Frugal Engineering*, Design Science and Innovation,
https://doi.org/10.1007/978-981-99-9700-8_6

using materials in excess of their requirements for achieving safety in the functioning of their products. This excess material or padding is provided to address the various uncertainties envisaged during the design process including scenarios of overloading; accuracy of theoretical models; materials and manufacturing; to name a few (Carper 2001) (Ullman 2003). Consequently, conservative designs inevitably consume more raw materials (Moynihan and Allwood 2014) (Mufti et al. 2005) and other resources to achieve their functionality with this excess translating into several products that could be fabricated with an eye on economy. In addition to irreversible draining of limited stocks of resources (Bardi 2014) (Gordon et al. 2006), the extra raw materials going into a conservative design would also translate into higher energy consumption and the concomitant emission of green house gases (GHG) (Moynihan and Allwood 2014) from activities needed to procure raw materials in a specified form (IPCC 2006). Other than planetary impact, conservative designs could also translate into higher costs (Duncan 2000) (Mufti et al. 2005). An instance of this all-around wastage is seen in the large tonnage of steel used in the construction industry due to selection of standardized components (Wise et al. 2013a). Another example is civil structures that continue to be conservatively designed and later fabricated to suppress failures (Moynihan and Allwood 2014) (Mufti et al. 2005). Therefore, the urge to being conservative in modern design runs counter to the need of our time (Bardi 2014) (Gordon et al. 2006). In particular, planetary crises (Steffen et al. 2011) compounded by an increasing population warrant economic utilization of earth's resources. Therefore, a revision in the design process is in order to account for the realities of this epoch. The prominence of engineering in the global economy (NAE 2002) makes such a change in design imperative for achieving better standards of living while seeking solutions for tackling planetary crises.

Hitherto, there have been several instances of streamlining design for good performance, e.g., designs based on minimum material or light weight constraint (Moynihan and Allwood 2014) and lower factors of safety (Beeby and Jackson 2016) (Duncan 2000); the Stradivari violin and; the Velodrome for the 2012 Olympic Games in London, UK. (Wise et al. 2013a). In recent years, the advent in emerging economies of low-cost frugal innovations with a no-frills structure, whose realization is constrained partly by scarcer material resources (Economist 2010), has been also roped into the private sector epitomized by companies like General Electric of the United States. In fact, material conservation has played a partial role in motivating the proposition of a National Network for Manufacturing Innovation (NNMI) in the US to counter the threat, among others, of low-cost manufacturers from emerging economies (US 2015) (Nature 2014).

The urge to design products for all round sustainable development has led to the advent of methodologies such as design for circularity (Ghisellini et al. 2016), design for dematerialization (Ewijk and Stegemann 2014) and product/service-system design (Vezzoli et al. 2015). These approaches, which are rooted in sustainability, conserve materials by reusing, recycling and reduction mechanisms; controlling material throughput through each stage of product life cycle; and offering a mix of products and services that cater sustainably to the needs of society; respectively. A widely popular approach is ecodesign that focuses on reducing environmental impact throughout the product life cycle comprising manufacturing, use and end of life (Roozenburg and Eekels 1995) (Rossi et al. 2016). The holistic nature of ecodesign subsumes the above design methodologies with material conservation being one of its features. However, ecodesign has had restraints in its real-time employment (Brones and de Carvalho 2015) not least due to the lack of metrics that can readily quantify its efficacy. Hence a metric combining classical design with frugality could be used in supporting and also quantifying either the methodologies of ecodesign or classical design with emphasis on material savings. It should be noted that the term classical refers to design traditionally based on the utilization of principles in strength-of-materials, mechanics and materials science for achieving a suitable factor of safety. This categorization has been adopted to distinguish and also facilitate development of the proposed frugality factor that is built on the classical model.

Therefore, there is a need for a suitable metric that besides donning the role of traditional *factor of safety* also quantifies material savings in various aspects of product cycle, including design, manufacturing and end of life and, whose widespread usage in design would encourage engineers and product developers to conserve resources. Although Life Cycle Analysis (LCA) is a vital tool (Bonou et al. 2016) for assessing environmental impacts of products (Consoli et al. 1993), it does not lend itself to a simple representation with the output being a strict function of accurate input data (Telenko et al. 2016). This chapter expounds on the development of the *factor of frugality*—an extension of the classical factor of safety—for quantifying the thrifty use of raw materials going into a product designed for quality performance and also safety.

6.3 Factors Influencing Frugal Design

Although classical principles are responsible for the bulk of "padded-designs", they can be used in the context of minimizing resource consumption. Such a change can be effected through schemes rooted in design theories both old or classical and new. The designer could directly apply such schemes individually or in combination to the problem at hand. Consequently, some schemes aiding frugality in consumption of raw materials are described in the subsections to follow. The content of these subsections underscore the possibility of conserving raw materials through judicious design principles. The designer needs to hew to the principles of these schemes, in the design phase, while leaving room for trade offs arising from parameter-optimization of individual parts and/or assembly of parts into a frugal system.

6.3.1 Rigorous Design Process

The padding in classical design could be minimized by strictly adhering to the loadings, vibrations and other vital parameters actually seen by a product during its working under real-time conditions. In recent years, Rolls Royce Inc (http://www.rolls-royce.com/about/our-technology/enabling-technologies/engine-health-management.aspx#sense) and General Electric (Economist 2015) have been using engine-health-management to gather real-time data for various critical parameters of jet engines. The loads, mechanical or otherwise, acting on a structure should be monitored real time to collect long-term data for informing the design process, thereby avoiding extrapolation and also lowering the uncertainty related to load sequence (Boller and Buderath 2007) (Branch 1976). The design should also be based on models giving accurate predictions; raw materials tested rigorously for their properties and; accurate forecasting of scenarios with a high likelihood of conditions such as overloading during the working of the product. Both predictions and forecasting should be based on well-tested mathematical models giving accurate predictions of critical parameters and stringent statistical analyses of data giving higher confidence levels. In this regard, the latest theories and/or data supporting lower uncertainty should be utilized in the design process. Furthermore, rigor in design should also involve statistically limiting the design parameters to narrow ranges accompanied by high confidence. The standardization-of-parts should be updated for allowing a wider range in sizes, thereby lowering the uncertainty involved in selecting the closest size available for a given component. Even non-standard components, whose fabrication is rendered difficult due to their optimal shapes, should be pursued due to their potential for significant weight reduction (Allwood et al. 2012).

Overall, a rigorous design procedure would minimize the uncertainties in classical design, thereby reducing the *factor of safety*. The lower *factor of safety* above unity—since one reflects working at failure load—would lead to minimal resource consumption that is commensurate to the actual performance of a product in the real world. Therefore, conservative designs in this effort are those that possess factors of safety above 1.5. Although this threshold is stringent when compared to the value reported by Otto and Antonsson (1991), its requirement is justified to emphasize both streamlining of design and robust performance for all round sustainable development. Other than aircrafts (Shanley 1962) (Norton 2006) (Torenbeek 2013), where a *factor of safety* of 1.5 is employed to reduce weight, stringent design processes are seldom applied in other sectors, not least, due to the higher costs encountered in their execution.

The advent of big data and advancement of affordable science and technology (Vincenti 1990) will eventually bring these rigorous procedures within the easy reach of other sectors. However, frugal products warrant rigorous testing under various scenarios of their usage to avoid failure in critical sectors such as aerospace, healthcare and others having significant bearing on human life. In fact, "frugalization" of critical sectors should be avoided in the absence of rigorous design procedures.

6.3.2 Alternative Design

Designers should envisage "simple" products that can be synthesized with a minimal number of components and/or features for minimizing resource consumption.

Apart from these no-frills products, alternative design forms, such as the onionskin model reported recently should also be made avail of. The onionskin model consists of longer lasting components located at the "core" with the "skin" comprising parts requiring frequent upgrades (Allwood et al. 2012). Such a modular approach facilitates segregation of parts belonging to the core and skin for ease of both upgrades and disassembly (Allwood et al. 2012). Another alternative design reported recently involves "folding" of components to build products with a high strength to weight ratio (Wood 2014). Altogether, simple alternative designs should be considered during the conceptual stage for prospective savings in raw materials when the product is realized.

6.3.3 Modern Manufacturing Techniques

Classical design leaves extra material for arriving at specified tolerances and surface finish during the fabrication of a part. In recent years, manufacturing has advanced to the point where a net shape can be created out of minimal excess material (Linton and Walsh 2003). Therefore, effective use of existing- and modern-fabrication techniques for skimming lesser material while arriving at the correct shape will lead to significant savings in raw materials. Besides processes, the streamlining of assembly operations has been shown to have a bearing on material savings. This is because products designed through Design For Assembly (DFA) typically have lower numbers of parts, connections and complex features to save on assembly time and hence have lower weight due to the concomitant savings in materials (Boothroyd et al. 1994). Accordingly, old- and new- manufacturing processes together with assembly operations should be investigated during the design stage to arrive at a part fabricated out of minimal amounts of raw materials (Boothroyd et al. 1994).

Other than approaching net shapes, manufacturing processes should also be utilized to impart beneficial features such as ultrafine-grained microstructures and proper sub-surface residual stresses for enhancing product life and hence saving materials (Valiev et al. 2000) (Withers 2007).

6.3.4 Mimicking Nature

The designer should enlist principles of Biomimetics where possible to save on material resources. Biomimetics entails adoption of design principles found in nature to the field of engineering thereby bringing the frugality and efficacy of nature's designs to human-made products (Ball 2001). Examples include design and fabrication of aircraft wings and wind turbine blades with whale-based tubercles to minimize drag for improved performance with a lower-weight-structure (Bhushan 2009). The design of the 1972 Olympic stadium in Munich was based on the principle of minimal surface area of membrane structures under surface tension, which translated into economy in weight (Ball 2009).

6.3.5 *Modern Materials*

The constraint on design due to dwindling resources could be met by focusing on resources themselves. The use of high-strength materials, including ultrafine- grained (UFG) (Valiev et al. 2000), could result in significant weight reductions vis-à-vis other relatively low-strength materials (Allwood et al. 2012). The principles of Biomimetics should be utilized to create low-weight materials possessing structure varying with scale that confers excellent stiffness and strength to the material concerned (Barthelat 2007) (Wu et al. 2014). Functionally graded materials could also be harnessed for achieving weight reductions in product design (Birman and Byrd 2007).

Even material testing should be based on samples consuming minimal amounts of raw material. In this regard, the crystalline sponges developed recently (Stallforth and Clardy 2013) offer the possibility of extracting molecular structures using X-ray crystallography of "nanograms" of sample material.

6.3.5.1 Exotic Materials

In recent years, materials such as Wood have been researched upon for their potential as a construction material (Mohanty et al. 2021; Song et al. 2018). Wood has been shown to possess good formability especially when hydrated (Mohanty et al. 2021) and this can be used to advantage in forming plates, spoons and other culinary articles for the food industry as shown in Fig. 6.1. Such Wood-based products have significantly lower embodied energy than plastics which are a scourge of environmental pollution. Besides, the Wood in these food objects is inexpensive and can be readily bio-degraded for reuse in other products. Wood has also been used as a construction material for buildings—see Fig. 6.2—and its potential as a structural material has been clearly documented by Song et al. (2018).

6.3.6 *Salvaging End-of-Life Components Through 4R Mechanisms*

In recent years, the cradle-to-cradle concept of sustainability has encouraged remanufacturing of discarded parts and "upcycling" of materials wherein raw materials going into a product are reused after that product's end of life (Hoornweg et al. 2013) (Bjørn and Hauschild 2013). The upcycling is made possible through melting or any other process that facilitates recovery of major portion of materials going into a product. Therefore, ideally, existing materials such as metals and their alloys could be used endlessly within the same or a different family of products (Wise et al. 2013b). Besides upcycling and remanufacturing, organic materials are being created, some of which are edible and others nourish various applications at end of life (Wise et al. 2013b).

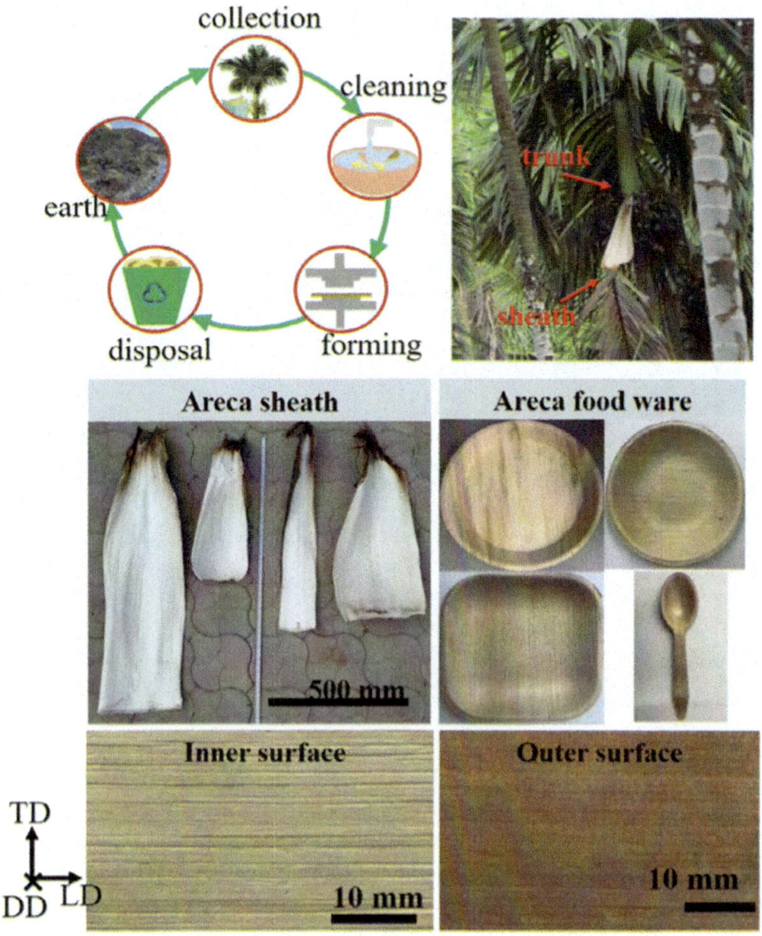

Fig. 6.1 Culinary objects created by forming of Areca palm sheath (Mohanty et al. 2021)

Conventional design principles could be followed for realizing products that are partially/completely built with parts from *end-of-life* (EOL) systems. An EOL system refers to a product that is past its prime and therefore no longer in use. Therefore, any such EOL system could be disassembled and, ideally, all of its parts utilized in other designs. The EOL parts could be utilized in a new design belonging to the same family or to a distinctly different system but compatible with the specifications of its design. In general, compatibility of parts from an EOL system can be improved by remanufacturing to facilitate its adaptation to a given product. Besides remanufacturing, EOL parts can also be salvaged through reuse, recovery and recycling, all of which comprise the 4R mechanisms (Seliger 2007). The use of EOL systems economizes resource consumption in classical design while also adhering to the cradle-to-cradle concept of sustainability. The end-of-life vehicle (ELV) programs

Fig. 6.2 A building made out of timber! (Tollefson 2017)

are a case in point, which involve substantial recycling of parts and materials for reuse in the automotive industry and elsewhere (Minter 2013) (Edwards et al. 2006) (Seliger 2007). Reuse of old parts has also been shown to have potential in other industries including construction and household appliances (Allwood et al. 2012).

6.4 Factor of Frugality

A rigorous design would have the *factor of safety* gravitate closer to the unit value (~1.5—taken from aircraft design, (Shanley 1962)), thereby minimizing uncertainties and hence resource consumption. Alternatively, a design with an arbitrary *factor of safety* could be realized with materials and parts conserved through any of the schemes, in Sect. 6.2, save the one on rigorous design. Therefore, maximum economization of raw materials could be achieved by taking the *factor of safety* above and closer to one with emphasis on safety and also use as many material saving schemes as possible.

A note on uncertainties in general. Uncertainties resulting in dispersions of concerned quantities are a bane of any scientific activity. As mentioned in this chapter, uncertainties are the reason for adopting conservative design practices involving higher *factors of safety*. However, in recent years, the coming of *big data*; progress in computing resources; and availability of relatively accurate models have enabled accurate characterization of uncertainties in various facets of engineering. Consequently, the establishment of a *factor of frugality*, as a new *safety factor* to greatly minimize material consumption without sacrificing functionality, has been made possible.

Accordingly, the *factor of frugality* is a function of both the *factor of safety* and material saving schemes to quantify the thrift in a given design effectively. In other words, the numeric value of the *factor of frugality* would account for the material saved through rigors of classical design and, in addition, the material saved through alternative designs, salvaging-4R-mechanisms, modern manufacturing techniques, modern materials, biomimetic design principles and other schemes foreseeable in the future. Hence, the *factor of frugality* (F in F^S) is expressed as a sum of *factor of safety* (S), from classical design and, a parameter termed material saved (MS) that accounts for raw material conserved through any combination of schemes outlined in Sects. 6.3.2 to 6.3.6. The equations summarizing the proposed *factor of frugality* are given by,

$$F = S + MS \tag{6.1}$$

$$S = \frac{S_M}{S_W} \tag{6.2}$$

$$MS = \sum_{I=1}^{N} MS_I \tag{6.3}$$

where the first term on the right side of Eq. (6.1) is the traditional *factor of safety* (S), as given by Eq. (6.2), which is the ratio between the maximum value of a material property (S_M) and working value of the material property (S_W). Therefore, S_M could be the yield strength (σ_Y) or the critical stress intensity factor (K_C) of the material used in the design and S_W would be the stress (σ) or the stress intensity factor (K), respectively, encountered during the actual working of the design (Shigley and Mischke 1989). The second term is the material saved (MS), which is equal to the total savings realized through all or any combination of schemes outlined in Sects. 6.3.2 to 6.3.6 with MS_I in Eq. (6.3) denoting the material saved in the Ith scheme. Although MS should be calculated from Eq. (6.3) for new designs, the value of MS is also equal to the difference between F and S due to the relation given by Eq. (6.1). Therefore, F^S is a symbolic representation that by packing in both F and S for ready reference also serves to readily quantify MS.

The individual components of material saved, i.e., MS_I, are listed in Table 6.1 along with the terminology for their definition. An assumption underlying these definitions is that all components except MS_5 have data available from initial design

Table 6.1 Formulations for the components of MS. Both the *factor of safety* and material remain constant in all of the definitions listed below

S. no.	Scheme	Formulation for MS component	MSMaximum
1	Alternative designs (Sect. 6.3.2)	$MS_1 = \frac{(W_{BULKY\ DESIGN} - W_{ALTERNATIVE\ DESIGN})}{W_{BULKY\ DESIGN}}$	0.5
2	Modern manufacturing techniques (Sect. 6.3.3)	$MS_2 = \frac{(W_{OLD\ PROCESS} - W_{NEW\ PROCESS})}{W_{OLD\ PROCESS}}$	1
3	Mimicking nature (Sect. 6.3.4)	$MS_3 = \frac{(W_{TRADITIONAL\ DESIGN} - W_{BIOMIMETIC\ DESIGN})}{W_{TRADITIONAL\ DESIGN}}$	0.5
4	Modern materials (Sect. 6.3.5)	$MS_4 = \frac{(W_{LOW\text{-}STRENGTH\ MATERIAL} - W_{HIGH\text{-}STRENGTH\ MATERIAL})}{W_{LOW\text{-}STRENGTH\ MATERIAL}}$	0.5
5	Salvaging through 4R Mechanisms (Sect. 6.3.6)	$MS_5 = \frac{W_{SALVAGED}}{W_{TOTAL}}$	1

$W_{ALTERNATIVE\ DESIGN}$: Weight of the simple or alternative design in Sect. 6.3.2.
$W_{BULKY\ DESIGN}$: Weight of the conventional bulky design.
$W_{NEW\ PROCESS}$: Weight of excess material removed by an alternative manufacturing process in Sect. 6.3.3.
$W_{OLD\ PROCESS}$: Weight of excess material removed by a traditional manufacturing process.
$W_{TRADITIONAL\ DESIGN}$: Weight of a bulky conventional design in Sect. 6.3.4.
$W_{BIOMIMETIC\ DESIGN}$: Weight of the design realized through principles of Biomimetics.
$W_{LOW\text{-}STRENGTH\ MATERIAL}$: Weight of a design using a low-strength material in Sect. 6.3.5.
$W_{HIGH\text{-}STRENGTH\ MATERIAL}$: Weight of a design using a high-strength material.
$W_{SALVAGED}$: Weight of all components salvaged through 4R mechanisms in Sect. 6.3.6.
W_{TOTAL}: Weight of the complete product including salvaged components

iterations for arriving at their numbers. Another assumption underlying these definitions is that the various MS values are computed by comparing designs having the same *factor of safety* and made of same material. The individual contributions to MS are defined so as to give an upper value of 0.5 or 1 corresponding to the maximal impact of the scheme concerned. In other words, all components of MS, except MS_5 & MS_2, tend to a value of 0.5 as a designer strives to maximize corresponding schemes. As opposed to the 4R mechanisms and manufacturing schemes, where it is possible to conserve 100% and hence take their respective MS components to unity, a maximum value of 0.5 has been selected as a conservative estimate for the remaining schemes due to the implausibility of complete conservation. Accordingly, MS_1 of Table 6.1, which is defined as the quotient of difference in weight between a bulky- and an alternative-design to the weight of the bulky design, tends to a value of 0.5 with decreasing weight of the alternative design. The remaining definitions are similar except for the 4R mechanisms, where $W_{SALVAGED}$ denotes the weight of materials salvaged from EOL systems for a given total weight (W_{TOTAL}) of the product concerned. Therefore, MS_5 approaches unity, as $W_{SALVAGED}$ comprises greater proportions of W_{TOTAL} with a maximum value of one for a design built completely out of EOL components. It should be noted that material saving schemes and, their MS components, not covered in this effort could also be incorporated into the model for F^S through procedures similar to those used in generating the entries of Table 6.1.

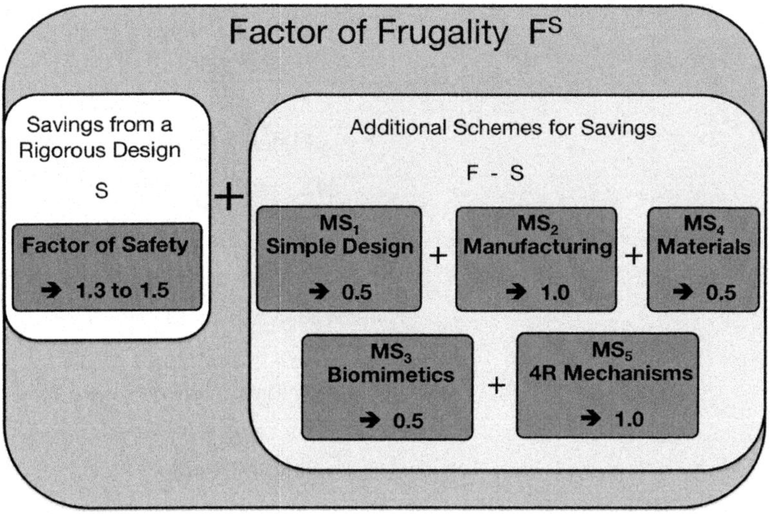

Fig. 6.3 Schematic for an optimal *factor of frugality*. Arrows signify approach of components to their maximal values

The safety factor captures the uncertainty of a given design while material saved accounts for the savings realized in opting for efficient scheme(s). For instance, both alternative and bulky versions mentioned in Sect. 6.3.2 could be designed rigorously with S being 1.5 that avoids material padding coming from higher factors of safety. However, it is MS_1 that quantifies the extra savings realized by opting for an alternative design, for an S of 1.5, with fewer number of parts and hence lesser consumption of raw materials. The same reasoning holds for other components of MS, each of which quantifies the savings realized by opting for the concerned scheme at a fixed value of S. This arrangement avoids redundancies in accounting for thrift by the components of F^S.

The F^S is a crucial metric that besides controlling design parameters also quantifies the savings achieved in the amounts of raw materials going into a product. A proper value of F^S is dictated by suitable values of its constituents, i.e., S and MS. Ideally, an S in the vicinity of one signifies a design, whose parameters have been determined rigorously with minimal uncertainties. A value of 0.5 for MS_1, MS_3 and MS_4 implies that the corresponding schemes have been utilized to their fullest practical extent. In contrast, a maximum value of one is possible for MS_2 and MS_5 since this would correspond to net shape fabrication without removal of excess material and a product comprising entirely of parts salvaged from EOL systems through 4R mechanisms, respectively. Moreover, the approach of individual components of MS to a value of 0.5 or 1, for maximal economy, synchronizes with S, whose approach to a value close to unity ($\sim$1.5) also signifies a thrifty design. Therefore, as depicted in Fig. 6.3, the maximum value for F in F^S, would be the aggregate of a number for S slightly higher than one and, an aggregate number coming from totaling the maximum values of

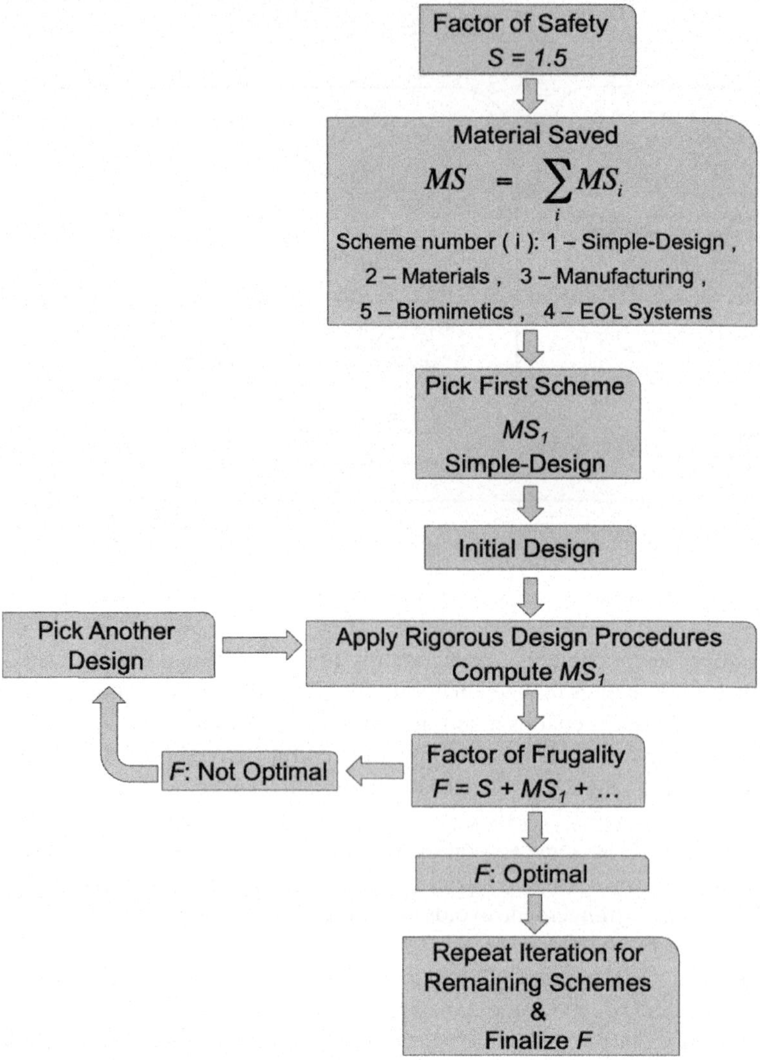

Fig. 6.4 Flow chart for computing the optimal *factor of frugality* (from Rao (2019))

the components of MS. For example, the maximum value for F^S is $5^{1.5}$ where F is 5 when S is taken to be 1.5 and maximum values employed for the five MS components shown in Fig. 6.3. A flow-chart is shown in Fig. 6.4 to help determine the *factor of frugality* for a given problem.

Equations (6.1), (6.2), (6.3) and, the definitions of Table 6.1, is a first generation model for F^S, whose material saving schemes are easily quantifiable. The model for F^S should be applied to the design of individual components of a product. These

Table 6.2 Efficacy of the *factor of frugality*. (The numbers listed in this Table are not specific to any real-time cases. These numbers have been created for expounding the workings of the *factor of frugality*. Maximum values of 1 and 0.5 have been assumed for MS_2 & MS_5 and MS_1, MS_3 & MS_4, respectively. For case 3 a maximum value of 0.5 has been assumed for each of the five extra components of MS.)

S. No.	Factor of frugality (**F_S**)	Factor of safety (**S**)	No of material saving schemes	Material saved MS (F – **S**)
1	7.5^4	4	5	3.5
2	$4.8^{1.3}$	1.3	5	3.5
3	$7.3^{1.3}$	1.3	10	6
4	4^4	4	0	0

individually designed parts will subsequently be assembled into a composite frugal product. This effort stops short of quantifying other aspects of MS such as material saved over the longer term by imparting apt textures and residual stresses through suitable manufacturing processes. More of these subtle savings in materials through relevant schemes should be characterized in futuristic efforts.

6.5 Computing *Factor of Frugality*

The efficacy of F^S in quantifying frugality of material consumption is brought out by the hypothetical cases of Table 6.2. It should be noted that all cases, except 3, use a MS value of 3.5 based on maximum values of components shown in Fig. 6.3. Case 3 uses five extra arbitrary schemes with each additional scheme contributing a maximum value of 0.5, which takes the MS value from 3.5 to 6.

The higher the difference F – S the higher the amount of material saved through frugal schemes in excess of savings coming from a given S. This is borne out by cases 2 and 3 of Table 6.2, whose extra savings through MS, correspond to values of 3.5 and 6 respectively, for an S value of 1.3. Although cases 1 and 2 have identical MS values for different values of S, i.e., 4 and 1.3 respectively, the total savings is optimal only for case 2. Even though case 1 has a higher F value, it has an S value of 4 that indicates higher uncertainty and hence more material wastage vis-à-vis the S value of 1.3 in case 2. Case 3 also highlights the potential for higher MS and hence a higher F number through additional arbitrary schemes realizable in future. The absence of any material saving schemes in case 4 makes the *factor of frugality* same as *factor of safety*. It should be noted that bulk of current designs following the classical approach are typified by case 4. Therefore, just following suitable material saving schemes would improve the MS and hence F numbers for these designs until further improvement in F is sought through the realization of tighter S values—whose realization could be abetted through advances in big data and also progress-in-technology.

6.5.1 Example: Frugal Design of a Shaft

An example on design of a shaft, as reported by Ugural (2015), has been selected in this study for exemplifying the proposed *factor of frugality*. The focus on shaft is justified by its significance as one of the fundamental elements in mechanical design that is widely used in applications requiring transmission of power and/or conversion of motion from linear to rotary and vice versa.

Table 6.3 lists results from both classical design, taken as the baseline for comparison and, the frugality approach—implemented through three of the schemes outlined in Sect. 6.3—for a shaft made of steel transmitting 500 kW of power at 1200 rpm. A *factor of safety* of 1.5 and relevant values for the strength and shear modulus of steel have been selected from Ugural (2015). The diameters reported here and, also listed in Table 6.3, are rounded to the nearest whole number for clarity. The first entry in Table 6.3, corresponding to baseline design, lists a solid steel shaft of diameter (D) of 46 mm that satisfies requirements of pure torsion during power transmission. The remaining entries pertain to frugality approach with the second entry on *alternative design* (Sect. 6.3.2) listing equivalent results for a hollow shaft, whose outside diameter (D) is 1.25 times the inside diameter (d). This results in a weight reduction of 50%, or MS_1 value of 0.5, which is reflected in the F^S value of $2^{1.5}$. The third entry refers to the use of a suitable manufacturing technique for producing the hollow shaft. The production of hollow shaft by indirect extrusion, as opposed to machining, results in an MS_2 of 1 as computed from entry 2 of Table 6.1. This value of MS_2 reflects significant savings from not having to machine the internal diameter out of a solid shaft, while taking insignificant finishing cuts on the inside and outside surfaces of both the extruded- and machined-shaft. The final entry lists the salvaging of such an extruded hollow shaft from an EOL system and accordingly adds 1, or $MS_5 = 1$ (obtained from entry 5 of Table 6.1), to the *factor of frugality*. Figure 6.5 illustrates the various outcomes at different stages of the frugal design process listed in Table 6.3.

Each of the individual material saving schemes adds up to give a *factor of frugality* of 4 with the representation of $4^{1.5}$. The *classical* approach yields a baseline design with an F^S value of $1.5^{1.5}$. In other words, the baseline design conserves material by adhering to a *factor of safety* of 1.5 without involving any of the additional conservation schemes. In contrast, each of the other schemes based on an alternative design, manufacturing and salvaging, respectively, lead to a cumulative improvement of the F value by 167%. This improvement in the *factor of frugality* attests to its significant potential for quantifying conservation of material resources and hence making the *classical factor of safety* stronger.

Other elements of mechanical design such as plates, gears, beams, springs etc. can also be similarly designed and fabricated for frugality. These individual *frugal* elements will get assembled into a composite whole to form a given *frugal*-product. In particular, each of these frugally designed components will typically be assembled into a *frugal* product (or *frugal innovation*). It should be noted that loads arising from interaction between individual frugal parts after assembly should be foreseen by the

Table 6.3 Frugal design of a shaft. Material is Steel (Shear strength = 300 MPa & Shear modulus = 80 GPa) Power transmitted = 500 kW, Rotational speed = 1200 rpm, *Factor of Safety* = 1.5

No.	Approach	Factor of safety (S)	Conservation schemes			Factor of frugality (F)	Outcome
			Simple design (MS_1)	Manufacturing (MS_2)	4R mechanisms (MS_5)		
1	Classical (Basic)	1.5	0	0	0	$1.5^{1.5}$	Solid Shaft
2	Frugal	1.5	0.5	0	0	$2^{1.5}$	Hollow shaft
3	Frugal	1.5	0.5	1	0	$3^{1.5}$	Tube extrusion
4	Frugal	1.5	0.5	1	1	$4^{1.5}$	Salvage extruded tube

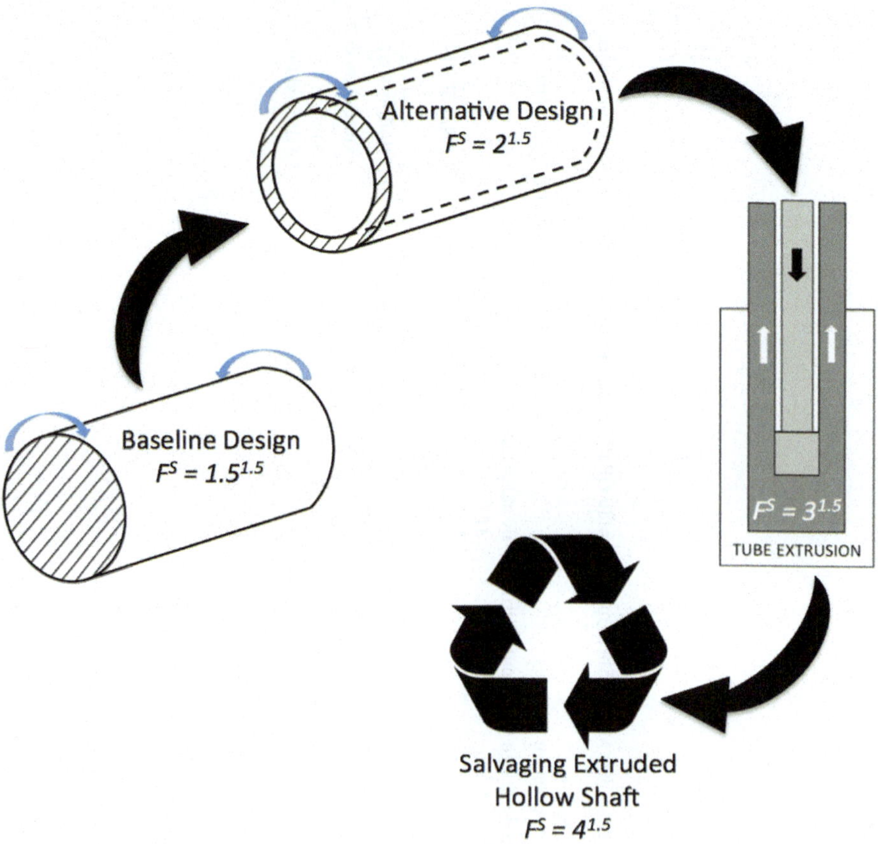

Fig. 6.5 Outcomes in the frugal design of a shaft. F^S values show the accompanying improvement in frugality

designer and used as commensurate constraints on the concerned parts during their frugal design.

6.5.2 *More Insight on Frugal Shaft*

A lower value of S typically results in a design operating closer to the failure condition with S = 1 denoting working stress equated to the failure stress. As a result, the use of a low S value should be accompanied by rigorous design procedures as in the aircraft industry where a S value of 1.3 to 1.5 (Shanley 1962) (Norton 2006) (Torenbeek 2013) is typically used for improving fuel efficiency. However, a lower *factor of safety* in design also gives the benefit of consumption of lesser amounts of materials

Table 6.4 Overloading in the classical design of solid shaft

No	Factor of safety (S)	Savings in weight (%)	Baseline torque (N-m)	Maximum torque (N-m)	Overload (%)
1	1.5	37	3979	5970	50
2	3	–	3979	11940	200

for realization of a product. *Frugal innovations*, either of the grassroots or advanced type with a no-frills structure, typically consume minimal amounts of material and hence fall into the low *safety factor* category and could thus be vulnerable to failure under nominal and/or even marginal conditions of overloading.

An instance of designing a shaft with a lower S has been demonstrated in this chapter. The shaft is a workhorse for transmitting power and/or conversion of rotary motion to linear and vice versa. The shaft has also been selected because its design brings out the features critical to robust functioning of frugal products in general. Table 6.4 lists results from shaft design with extrema in S of 1.5 and 3. The case of S=1.5 has already been covered in this chapter. The results listed in Table 6.4 clearly show that savings in weight are accompanied by reduced margins for overloading as S goes down to a value of 1.5. The overloading entries are calculated from the baseline- and maximum-torque, which correspond to the basic design and torque resulting in incipient plastic deformation, respectively.

A *factor of safety* of 3 in Table 6.4 results in a design that can handle an overload–in excess of rated torque–of 200% vis-à-vis 50% tolerated by the baseline design with S of 1.5. The amount of overload that can be withstood by a design becomes even lower for S tending closer to unit value with S = 1 signifying incipient plastic deformation. Therefore, such weaker margins in overloads for a very low S would possibly lead to the early onset of failure in the shaft and *frugal* products in general. However, the design for S = 3 uses materials in excess by weight of 37% when compared to baseline.

6.5.3 *Why Frugal Shaft is Better*

Table 6.5 lists results of designing the shaft through frugal approach that employs the *factor of frugality* without iterations. The first entry corresponds to classical design of a solid shaft (first entry of Table 6.4) whose *factor of frugality* is same as the S of 1.5. The second entry, illustrated in Fig. 6.5, lists results of the *frugal* approach for an S of 1.5 where extra savings from selecting a hollow shaft by classical design; extrusion as opposed to machining of such a hollow shaft; and salvaging such an extruded hollow shaft from an EOL system; have been quantified and added to give a F^S of $4^{1.5}$. It should be noted that the baseline case could be the hollow shaft itself but starting with a solid shaft covers the general case and also brings out the additional weight savings realized by opting for a hollow part. Moreover, the *frugal* approach

Table 6.5 Overloading capacity of frugal shaft

No	Design approach(S)	Material saved			Factor of safety(S)	Factor of frugality (F)	Overload (%)
		Alternative design (MS_1)	Manufacturing (MS_2)	Salvaging (MS_5)			
1	*Baseline*	1.5	–	–	–	$1.5^{1.5}$	50
2	*Frugal*	1.5	0.5	1	1	$4^{1.5}$	50

should be applied to different designs for a given product and the optimal design being the one with the highest F value for the given application, see Fig. 6.4. Only three of the five MS factors of Table 6.1 are utilized here because of their relevance to shaft design. Table 6.5 brings out the maximum savings in material possible for the shaft by combining the low *safety factor* design with additional savings realizable through the approach of this chapter.

Frugal innovations fall in the category of a low safety factor design wherein material savings are achieved at the expense of vulnerability to failure under various conditions of loading. Such vulnerability can impair the functionality of *frugal* products, irrespective of the grassroots or advanced types, with serious consequences especially in critical sectors such as healthcare. In fact, grassroots innovations could be vulnerable even within the domain of nominal working conditions due to their makeshift nature that excludes application of any design methodologies. In particular, any of the following and, other reasons, individually or in combination, could lead to premature failure of a makeshift frugal innovation: intuitive economizing of material usage that pays no scientific attention to weak spots; impromptu medley of EOL parts, whose integrity individually has not been vetted; and improper assembly of parts, to name a few.

However, a low S design is used to advantage in the *frugal* approach. The lower *factor of safety* of 1.5 is selected for the ensuing attendant rigorous design procedures. The rigor in design is warranted to both avoid material padding to offset design related uncertainties and achieve robust functionality under nominal working conditions with a possibility of even sustaining some overloads–as inferred from the design showcased in Table 6.5 where the *frugal* or hollow shaft is capable of sustaining a 50% excess torsional-load. It should be noted that both the hollow and solid shafts are capable of sustaining a 50% overload, see Tables 6.5 and 6.4 respectively. Other than sustaining overloads, an instance of a frugally designed product having better load bearing abilities vis-à-vis classical one is evidenced by the hollow shaft in Table 6.5 that has improved sectional properties (Timoshenko and Young 1968) for enhanced structural rigidity when compared to a solid shaft designed under the same set of conditions for a S of 1.5. Therefore, products scientifically designed by the *frugal* approach are not necessarily *frugal* in their performance. In addition to economizing from the baseline step of the *frugal* approach, other schemes rooted in sustainability are also utilized to skim as much material as possible for a given product. In the shaft example covered in this work these schemes involved the use of hollow feature, as an alternative design, to minimize consumption of raw materials; manufacturing the hollow shaft with reduced waste; and salvaging such an extruded hollow shaft. Overall, the use of these extraneous material saving schemes on top of a low S design have improved the *factor of frugality* by 167% in going from 1.5 to 4.

Although the *factor of safety* approach can also be applied to designing such innovations, the *frugal* approach brings out the additional material saving schemes that are not accounted for by the classical design process. In particular, material saved through schemes such as salvaging from EOL systems and use of an apt manufacturing technique are extraneous and hence completely absent in the classical design approach. Ideally, the *factor of frugality* sums up savings from all possible sources to truly quantify the *frugal* use of materials while aiming for robust functionality of product.

Even though classical design principles have been utilized in arriving at F^S, the *frugality factor* has been established to formally encourage and also ease adoption of green practices by *frugal*- and non-*frugal*-designers alike for all round sustainable development. Therefore, low *safety factor* coupled with rigorous design; extraneous material saving schemes; and low-cost are significant features of *frugal engineering* (and design in particular) that would beckon designers and engineers to go *frugal* especially when humankind needs to simultaneously fight *climate change* and uplift living standards of society at large. In this regard, Eq. (6.1) can be applied to systematically build *frugal* products from scratch and also attribute a F^S score to rank their credentials for frugality or judicious use of resources. Such ranking and subsequent labeling of products with F^S scores would, as a tool of ecodesign, facilitate efforts for progressing in all round sustainable development.

6.6 Factor of Frugality is Universal

It has been mentioned several times that the *safety factor* has to be as low as possible for a proper *factor of frugality*. However, this does not limit the applicability of F^S even to applications with a higher S. An S of $\sim$5 is common in the nuclear industry where human safety warrants such higher S values. Even in such designs F^S can be utilized to advantage by making use of the MS parameters to save as much as possible in schemes extraneous to high-S design. In fact, research could aid in reducing S in addition to savings from MS.

Therefore, there is no single low S value encompassing all products in all sectors. The astute designer will have to use F^S judiciously relative to the application and its context at hand. Hence, the measure of F^S proposed in this chapter is universal, i.e., applicable to any product!, with research helping in progressively reducing S, MS and hence F with time. Newer methodologies in design, manufacturing and materials such as generative design, topology optimization and Wood respectively are all within the ambit of the *factor of frugality*. Therefore, the factor of frugality approach is a general approach whose breadth of choice for the designs and their development (that includes materials, manufacturing, salvaging) is wide enough to consider traditional and newer research findings that keep coming up endlessly.

6.7 Factor of Frugality Versus Reliability Approach

The *reliability* based approach is the other methodology popular in traditional design. *Reliability* is defined as the probability of functioning of a product/part without failure in a certain window of time. That is *reliability* (R) is given by (Shigley and Mischke 1989),

$$R = 1 - P_{\mathrm{F}} \tag{6.4}$$

where P_F is the probability of failure according to frequentist interpretation of probability theory. The *reliability* approach to design is stochastic in nature. In other words, this approach to design needs proper probability distributions of strength, stress and other relevant parameters affecting a given problem. Therefore, its success is a critical factor of the data used to characterize these distributions. Although in this age of big data, statistical techniques are de rigueur, their use entails encountering more uncertainties which have to be rigorously tackled for the *frugal design approach.*

In contrast, the *factor of frugality* gives a single number obtained through rigorous design procedures with a keen eye on constraining resources and costs. The *factor of frugality* is an extension of the traditional *safety factor*, which it subsumes, and as such is easier to apply like the ubiquitous *factor of safety*. Moreover, use of the latest and most accurate models and also data significantly reduce uncertainties in the *factor of frugality* approach.

Future efforts should combine the *reliability* approach with that of *frugality factor*. An alternate route could also be modification of the *reliability* approach to account for frugality-based constraints.

6.8 F^S for Sustainable Development

The *factor of frugality* can also be applied inversely in the conceptual stage to simulate a proper mix of safe-and-rigorous-design and material saving mechanisms. The variation in F of F^S and its components should be thoroughly studied in the initial design stage, prior to fabrication, for realizing optimal savings in materials. Although S is subsumed in F^S, it will continue its role of making products both functionally robust and safe by presenting itself in the relevant formulas of engineering design. Since design is an iterative process, the *factor of frugality* and its constituents, in some cases, will have to be evaluated in finite iterations when weak spots are revealed during initial design cycles.

The *factor of frugality* would also aid in assessing the impact of a design on *climate change*. An optimal value of F^S would signify judicious use of raw materials that generally translates (Moynihan and Allwood, 2014) into lower GHG emissions. Therefore, frugality in resource consumption might also lead to a product, whose realization is a net saver of energy with a positive impact on climate change. However, a given design should be scrutinized for hotspots in its cradle-to-grave cycle, such as raw material extraction and disassembly, which could exhale larger amounts of GHGs and thus make the design a net emitter. Therefore, values of F^S could be scrutinized in tandem with relevant results from the *Life Cycle Analysis* (LCA). Such a combined study will aid in concluding definitively about the emission credentials of a given design. In fact, a single number from LCA could be appended to the *factor of frugality*, as in $F^S{}_{LCA}$, in a futuristic effort. Consequently, maximizing *factor of frugality* while sticking to the rigors of design could lead to better products that are environmentally benign. Therefore, adoption of F^S would revamp the *classical* design process and usher in a philosophy based on designing-for-frugality that stresses performance and economy in material utilization.

6.8.1 Diffusion of Factor of Frugality

The urgency to improve our environment through widespread sustainable development is warranted due to the following reasons. First, climate change is making it incumbent upon us to significantly lower our carbon footprint so as to leave a better planet to posterity. Second, the swelling ranks of people in lower- and middle-income segments around the world are aspiring for better living standards (Mekonnen and Hoekstra 2016) that are also affordable. In this regard, systematically designed *frugal innovations* offer the right mix of lower carbon footprint, robust functionality and lower price. However, excluding science by employing makeshift techniques to fabricate these innovations could result in premature product-failure that can impede the diffusion of these innovations and hence impair their significant potential for sustainable development. It should be noted that the *factor of frugality*, as given by Eq. (6.1), is built to account for *frugal* features originally seen in grassroots innovations and later observed in advanced products. Therefore, applying the frugal design approach to a grassroots concept will lead to a successful and systematic scientific-realization of a grassroots innovation. Other than grassroots ingenuity leading to the concept of *frugal* design, this is all the more important due to the significant impact grassroots innovations can make to the rich and poor alike in a sustainable economy when systematically designed and fabricated by leveraging scientific principles. The *factor of frugality* therefore is an apt ecodesign (Roozenburg and Eekels 1995; Rossi et al. 2016) measure for facilitating maximum savings in material resources and also quantifying it for any innovation qualifying as *frugal*.

Although creators of *advanced frugal innovations* can readily adopt the *frugal design approach* due to their possession of requisite wherewithal, grassroots innovators typically lack the skills and also resources to do so. Therefore, governmental and relevant non-governmental entities should set up subsidized programs to empower grassroots innovators with science including *frugal design* through academicians, scientists and other qualified workers. Public entities should shoulder the main responsibility of creating awareness and also informing *frugal* innovators in general about relevant institutions and associated scientific workers to contact for seeking design solutions. These services including actual design work should be offered free of cost to grassroots innovators. The private sector should also be encouraged to select and support, in terms of funding and knowledge, design work for promising grassroots innovations that could possess potential for commercial success. Overall, active involvement of public, private and other relevant entities in educating, funding and empowering grassroots *frugal* innovators with the underlying science, including *frugal design* methodology, should be encouraged.

6.8.2 Eco-efficiency

The *factor of frugality* is also an apt measure of eco-efficiency for all round sustainable development. The focus on economizing usage of raw materials aligns with the premise underlying eco-efficiency, i.e., minimize wastage (Montgomery 1997). In fact, products could be showcased for their frugality by displaying F^S values on their labels. A higher value for $F - S$ together with a suitable S would signify optimal savings in raw materials. In this regard, standardizing the procedure for evaluating F^S, including number of MS schemes and their maximal values, would facilitate comparison of F^S between products.

The close proximity between manufacturing and design is crucial for innovations (MIT 2013). In fact, exodus of many manufacturing facilities offshore has disrupted this proximity thereby impairing developed countries from harnessing manufacturing for lucrative innovations. In this regard, use of F^S, which explicitly accounts for both materials and manufacturing through MS, would facilitate quantification of improvements in design through schemes such as modern manufacturing thereby encouraging product development activities under one roof for innovating effectively in a globalized economy. Last but not least, F^S could be employed for both the systematic design of frugal innovations, which have become popular in recent years (Rao 2013) and, the study of their dynamics of interdependencies. In particular, use of F^S values as proxies for multitudes of frugal innovations could aid network studies seeking to bring out any instabilities arising from overcrowding of these innovations (Hellmann et al. 2016) (Helbing 2013).. In fact, the next chapter expounds on the complex networks of multitudes of interconnected *frugal* products, which will shed more light on the design of AFIs.

6.8.3 Factor of Frugality is the New Age Safety Factor

The philosophical underpinnings of a new *factor of safety* for tracking both safety and frugality in the ecodesign of a product, with a first generation model, have been presented in this chapter. The *factor of frugality* (F) subsumes the classical *factor of safety* (S) widely prevalent in engineering design. The representation F^S, for the new factor, teases apart contributions from both a rigorous design process (S) and various additional material saving schemes (F − S) to bring out the environmental credentials of a given product. The increasing value of the *factor of frugality* beyond 1.5, against a maximum of 5, signifies increasing savings in the resources consumed by a given product designed rigorously with a *factor of safety* of 1.5. The example of frugal shaft design has shown the potential to conserve material through a rigorous alternative design that is realized by salvaging an extruded hollow shaft from an EOL system. Even without a tighter *safety factor*, the use of material saving schemes built into the *factor of frugality* would result in realization of products with lesser wastage of raw materials. Other than quantifying the rigor in product design and also savings in raw materials, F^S can be used as a measure of eco-efficiency and proxy for network studies related to overcrowding of *frugal innovations*, to name a few of its applications. The *factor of frugality* can be applied to various industrial sectors, including aerospace or healthcare, after accounting for their specific constraints and complexity of design. These distinct sectors will get tuned to the rigor and frugality of the approach with time.

The use of the *factor of frugality* from here onwards could upend *classical* design philosophy and create products that are in sync with principles and policies for mitigating the impacts of climate change and resource scarcity. This is borne out by the usefulness of all the individual parameters going into the making of F^S. In products for which S might not hold sway, the MS schemes and related parameters could be used to monitor savings in resources and hence cost while rigorously developing the product. An instance of such an application will be related to paper-based AFIs that were reported earlier in this book. Table 6.6 below lists the features and brings out the advantages of using the *factor of frugality* vis-à-vis the classical *safety factor* in design from here onwards. It is noteworthy that even without a *frugal* mindset, the *frugal design approach* can revamp traditional design for saving resources, improving quality and lowering costs, all of which will uplift living standards and hence significantly influence *sustainable development*.

Descriptive Questions

1. What are the highest and lowest values of the *factor of safety* used in practice? Mention the industrial sectors corresponding to these values.
2. Why are *safety factors* high in certain industries?

Table 6.6 Features and, hence benefits, of *factor of frugality* vis-à-vis classical *safety factor*

No	Factor of frugality (**F_S**)	Factor safety (S)
1	F = S + MS	S
2	Composite factor with several constituents	Stand alone factor with no constituents
3	S is one of the constituents of F^S	Has no constituents
4	Other than S, F^S has constituents for material saved from schemes extraneous to traditional design	Supports only traditional design. No extraneous schemes to save more material
5	Design directly follows stages in product development such as manufacturing and salvaging	No direct dependence on stages of product development
6	Extraneous schemes account through MS for stages in product development such as manufacturing and salvaging	No direct dependence on stages of product development
7	Created to save material at every stage of product development	Created only for design stage of product development
8	Supports rigorous design practices	Flexible enough to accommodate even conservative or padded designs
9	Results only in streamlined design	Streamlined or bulky designs depending on its value
10	Excellent product functionality	Could compromise product functionality
11	Focus on lowering costs	No restraint on costs
12	Easy application to sustainable-design from scratch	Cumbersome application to sustainable-design from scratch
13	Applicable to any product of any sector	Applicable to any product of any sector

3. Can the *frugal design approach* be applied to industries using higher *safety factors*?
4. Can the *factor of frugality* be used effectively for industries using higher *safety factors*? Explain your answer in terms of the various terms appearing in Eq. (6.1).
5. Describe the steps involved in applying the *frugal design approach* to a given design problem.
6. Describe the steps involved in solving the design problem of question 5 in terms of traditional design approach.
7. Explain the differences in design outcomes of questions 5 and 6 corresponding to the *frugal* and traditional design approaches.
8. What are the pros and cons of both the *frugal* and traditional design approaches.
9. Select any traditional design problem of your choice and make a *frugal* version of it through the approach of this Chapter. ***Hint: Pick a simple component !! Subsequently attempt complicated products including those involving assembly of various parts.***
10. Explain the advantages of design based on *factor of frugality* when compared to the classical type.

11. The *factor of frugality* can be useful in designing a complete or portion of a product. Explain how this approach can be applied to cases where formal design techniques cannot be applied. **Hint: Use the MS parameters along with general concept underlying the *frugal design approach*.**

Problems

All the problems covered in standard *strength of materials* and *design* textbooks can be looked at from the prism of the *factor of frugality*. Here a few sample problems from Ugural (2015) have been adapted to indulge the reader into solving these problems using the *factor of frugality* approach. Please make realistic and plausible assumptions in quantifying the various schemes for the MS parameters appearing in Eqs. (6.1–6.3). The interested reader/designer should attempt, using the *factor of frugality* approach, solving more problems from the standard textbooks on Mechanical Design. **Hint: In all of these problems or any other, remember the assumptions underlying Eqs. (6.1) to (6.3) of this chapter. Especially the low *factor of safety*!**

1. A short steel pipe of *yield strength* 280 MPa is to support an axial compressive load of 1.2 MN with a certain *factor of safety* against yielding. Based on the *frugal design approach* determine the minimum inside radius required for this steel pipe.
2. A high-strength ASTM-A242 steel (Young's modulus = E=200 GPa) rod of length 6 m, used in a control mechanism, must carry a tensile load of 10 kN without exceeding its yield strength of 250 MPa (with a suitable *factor of safety*) and also not stretching more than 5 mm. Answer the following using the *factor of frugality* approach:
   a. What is the required diameter of the rod?
   b. What is the spring rate for this rod?
3. A pin-connected two-bar assembly or hoist is supported and loaded as shown in Fig. 6.6. Using the *factor of frugality* approach, determine the cross-sectional area of the round aluminum eye-bar AC and the square wood post BC. The required load is P = 50 kN. The strength of aluminum and wood are 480 and 60 MPa, respectively. It should be noted that the load acts in the plane of the hoist and weights of members are insignificant compared to the applied load. Also, friction in pin joints and the possibility of buckling in relevant members are ignored.
4. A two member pin-connected structure supports a concentrated load P at joint B as shown in Fig. 6.7. Calculate the largest load P that may be applied through the *factor of frugality* approach. The Young's modulus can be assumed to be 210 GPa. Also, assume that buckling occurs in the plane of the structure.
5. A steel column with an rectangular cross section (a $\times$ b) is fixed at the base and supported at the top, as shown in Fig. 6.8. The column must resist a load of 15

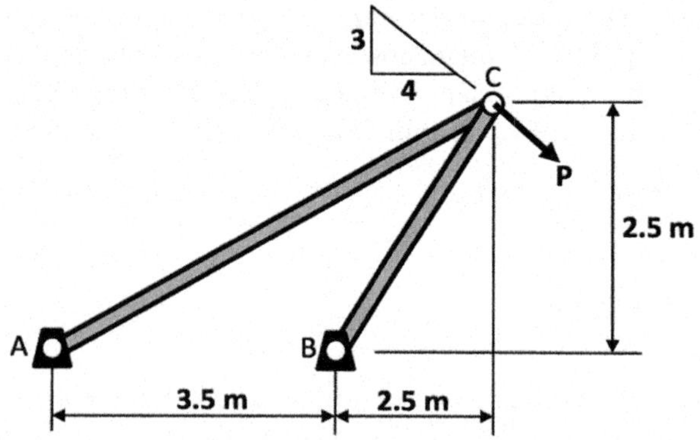

Fig. 6.6 Pin connected hoist example adapted from Ugural (2015)

kN as shown with a suitable *factor of safety* against buckling. What is the ratio of a/b for a *frugal* design against buckling? Design this *frugal* cross section for the column, using a Young's modulus of 200 GPa. Note that the restraints shown at the top of the column are such that only out-of-plane movement is allowed.

6. As shown in Fig. 6.9, a tensile link of thickness (t) with two fillets of r = 16 mm is subjected to a load fluctuating between P_{MIN} of 90 kNand P_{MAX} of 210 kN. Use a suitable *factor of safety* through *frugal design* for arriving at a value for the thickness. Use a D of 120 mm and a d of 80mm. Also, the link is made of steel with a *ultimate tensile strength* of 700 MPa. Assume unlimited life based on the Goodman criterion. Assume also that the fillets and adjacent surfaces are ground and a reliability of 99.9 % is desired.
7. Determine the diameter (d) of shaft AB shown in Fig. 6.10 with the maximum shear stress in each shaft being limited to 150 MPa. Use the frugality approach through a suitable *factor of safety* against shear.
8. A 22-tooth helical pinion has a normal pressure angle of 20°, a normal diametral pitch of 8 teeth/in., a face width of 2 in., and a helix angle of 25°. This pinion rotates at 1800 rpm and transmits 30 hp to a 40-tooth gear. Determine the lowest *factor of safety* based on bending strength, employing the Lewis equation. Use this *safety factor* and carry out a *factor of frugality* based design of the gearset. Use a fatigue stress-concentration factor of 1.5. Also, assume a suitable material for the pinion and gear in keeping with the tenets of the *frugal design approach.*
9. A disk of uniform thickness is used at 12,000 rpm as a rotating blade for cutting blocks of paper or thin plywood. The disk is mounted on a shaft of 1 in. radius and clamped, as shown in Fig. 6.11 below. The disk outside radius is taken as 15 in. This arrangement will have a certain *factor of safety* according to the maximum shear stress criterion. Is this the best *safety factor* from a *frugal design* perspective. Assume that the cutting forces are relatively small; speed is steady; and loading is considered static. The stress concentrations due to clamping and sharpening at the periphery are disregarded. Also, assume a suitable disk material from a *frugal design* perspective for the application concerned.

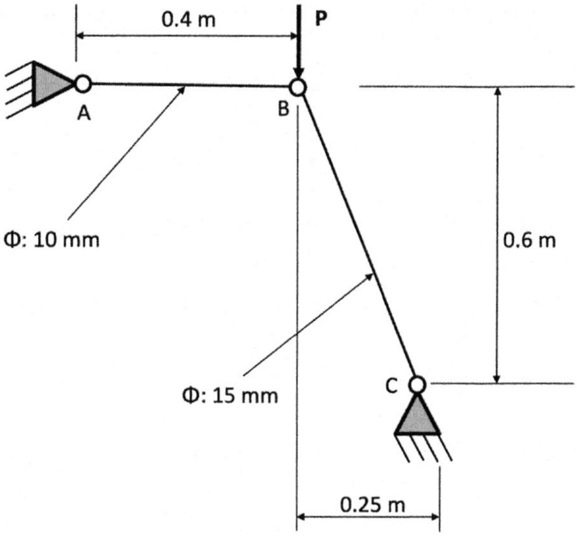

Fig. 6.7 Pin connected structure example adapted from Ugural (2015)

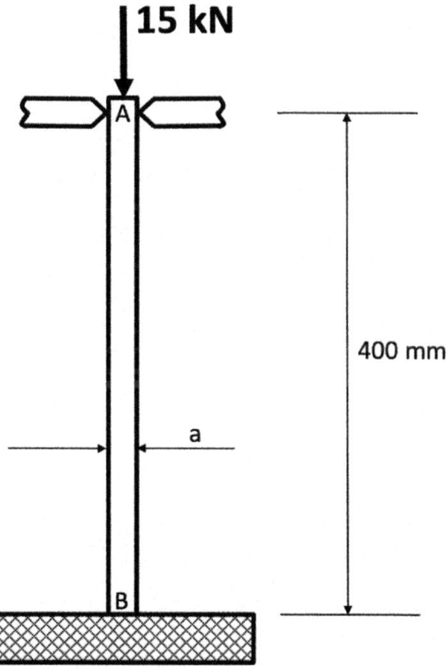

Fig. 6.8 Example on buckling of a rectangular column adapted from Ugural (2015)

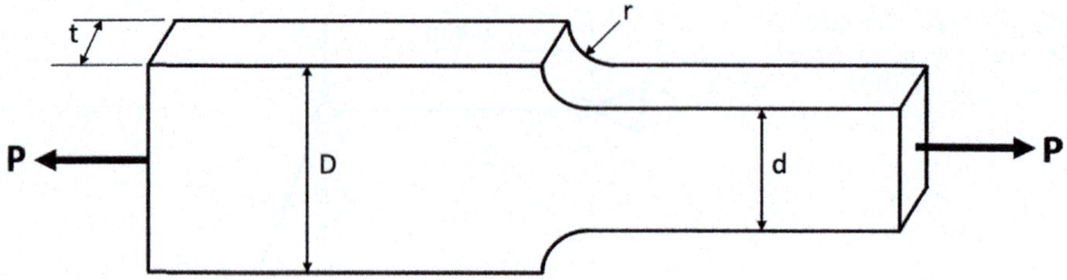

Fig. 6.9 Example on fatigue failure adapted from Ugural (2015)

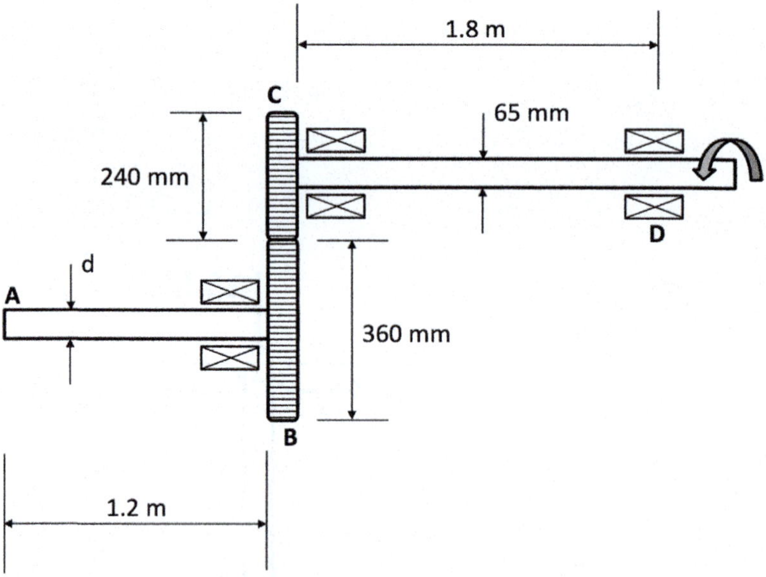

Fig. 6.10 Example on design against shear adapted from Ugural (2015)

Fig. 6.11 Example on rotating blade for paper cutting application adapted from (Ugural 2015)

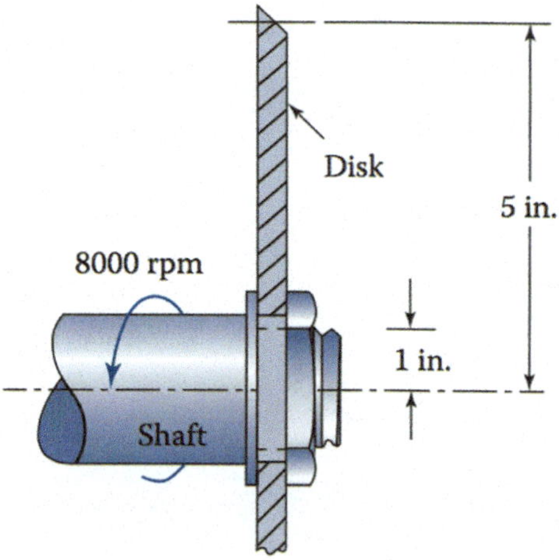

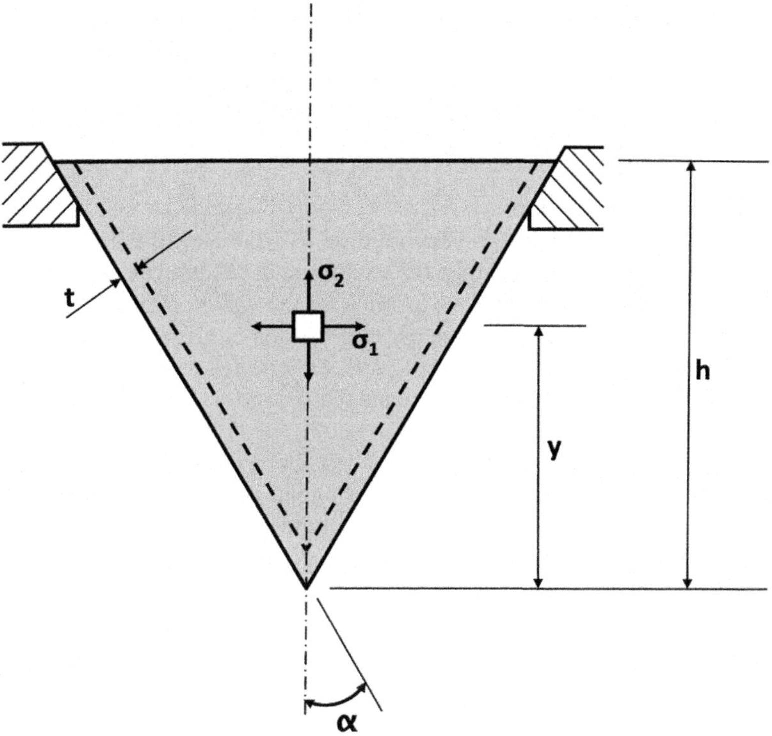

Fig. 6.12 Conical tank example adapted from (Ugural 2015)

10. A thin-walled cylindrical pressure vessel of *diameter* (D) is subjected to an *internal pressure* (P) varying continuously from (P_{MIN}) to (P_{MAX}). Determine the *thickness* (t) using the *factor of frugality* approach for an *ultimate strength* (S_U) and modified *endurance limit* (S_E). Use a D of 1.5 m; P_{MIN} of 0.8 MPa; P_{MAX} of 4 MPa; Yield strength of 300 MPa; S_U of 400 MPa and S_E of 150 MPa. Use the Goodman theories based on maximum normal stress for this problem and make plausible assumptions for material and *safety factor* in arriving at your solution.
11. A spherical pressure vessel of radius 2.5 ft is subjected to an internal pressure of 1.5 ksi. Determine the critical wall thickness (t), using the *factor of frugality* approach, and the corresponding diametral extension. A proper *safety factor* has to account for bursting failure. Assume a *ultimate strength* of 60 ksi, a Young's modulus of 30,000,000 psi and poisson's ratio of 0.3.
12. The *factor of frugality* approach is dependent on materials among other constituents of product development. Solve *problem 8* with an alternative material for designing a spherical pressure vessel thinner than the one obtained in the previous problem.

13. A large plate of width (2w) with (w = 60 mm) carries a uniformly distributed tensile force of 160 kN in the longitudinal direction with a *safety factor* that should be in accord with the *frugal design approach*. The plate has a central transverse crack that is (2a) long, with (a = 9 mm). The plate is made of titanium Ti-6Al-4V alloy. Calculate the thickness (t) required to:
    1. Resist yielding
    2. Prevent sudden fracture
14. A thin-walled conical tank is supported on its edge and filled with a liquid, as shown in Fig. 6.12. Accordingly, the vessel can be taken to be simply supported between top supports. Derive the formula for the wall thickness (t) based on the maximum shear stress and the energy of distortion failure theories. The vessel is made of suitable structural metal of *yield strength* (S_Y) and the *factor of safety* against yielding is (S). It should be noted that σ_1 and σ_2 are the *principal* stresses for the element shown.
15. Study the formula derived in *problem 14* to observe variation of **t** with **S**. Also, study variation of **t** against **S** for various engineering metallic materials. What combination of metal and *safety factor* would give a thickness value suitable for *frugal design approach*? Subsequently, compute the *factor of frugality* for *problem 14* by also considering relevant *material saved* schemes.

References

Allwood J, Cullen J, Carruth M, Cooper D, McBrien M, Milford R, Moynihan M, Patel A (2012) Sustainable materials: with both eyes open: chapter 12 using less metal by design; chapter 15 re-using metal components; and chapter 16 longer life products. UIT Cambridge, Cambridge, UK

Ball P (2001) Life's lessons in design. Nature 409:413–416

Ball P (2009) Shapes. Oxford University Press, New York, USA

Bardi U (2014) Extracted: how the quest for mineral wealth is plundering the planet. Chelsea Green Publishing, Vermont, USA

Barthelat F (2007) Biomimetics for next generation materials. Philos Trans Roy Soc A 365:2907–2919

Beeby A, Jackson P (2016) Partial safety factor for reinforcement. Structures 5:101–111

Bhushan B (2009) Biomimetics: lessons from nature-an overview. Philos Trans Roy Soc A 367:1445–1486

Birman V, Byrd L (2007) Modeling and analysis of functionally graded materials and structures. Appl Mech Rev 60:195–216

Bjørn A, Hauschild M (2013) Absolute versus relative environmental sustainability: what can the cradle-to-cradle and eco-efficiency concepts learn from each other? J Ind Ecol 17(2):321–332

Boller C, Buderath M (2007) Fatigue in aerostructures-where structural health monitoring can contribute to a complex subject. Philos Trans Roy Soc A 365:561–587

Bonou A, Skelton K, Olsen S (2016) Ecodesign framework for developing wind turbines. J Clean Prod 126:643–653

Boothroyd G, Dewhurst P, Knight W (1994) Product design for manufacturing and assembly. Marcel Dekker Inc, New York, USA

Branch G (1976) High temperature mechanical properties as design parameters. Philos Trans Roy Soc A 282:181–192

Brones F, de Carvalho M (2015) From 50 to 1: integrating literature toward a systemic ecodesign. J Clean Prod 96:44–57

Carper K (2001) Forensic engineering. CRC Press, New York, USA

Consoli F, Allen D, Bounstead I, Fava J, Franklin W, Jensen A, de Oude N, Parirish R, Perriman R, Postlethwaite D, Quay B, Seguin J, Vigon B (1993) Guidelines for life-cycle assessment: A code of practice., Technical report, Society of Environmental Toxicology and Chemistry (SETAC), Pensacola, FL, USA

Duncan J (2000) Factors of safety and reliability in geotechnical engineering. J Geotech Geoenviron Eng 126(4):307–316

Economist (2010) First break all the rules: the charms of frugal innovation

Economist (2015) Smart products, smart makers

Edwards C, Coates G, Leaney P, Rahimifard S (2006) Implications of the end-of life vehicles directive on the vehicle recovery sector. Proceedings of the institution of mechanical engineers part B: J Eng Manufact 220:1211–1216

Ewijk SV, Stegemann J (2014) Limitations of the waste hierarchy for achieving absolute reductions in material throughput. J Clean Prod 132

Ghisellini P, Cialani C, Ulgiati S (2016) A review on circular economy: the expected transition to a balanced interplay of environmental and economic systems. J Clean Prod 114:11

Gordon R, Bertram M, Graedel T (2006) Metal stocks and sustainability. Proceed Nat Acad Sci (PNAS) United States America 103(5):1209–1214

Helbing D (2013) Globally networked risks and how to respond. Nature 497:51–59

Hellmann F, Schultz P, Grabow C, Heitzig J, Kurths J (2016) Survivability of deterministic dynamical systems. Sci Rep 6:29654

Hoornweg D, Bhada-Tata P, Kennedy C (2013) Comment: Waste production must peak this century. Nature 502:615–617

IPCC (2006) Guidelines for national greenhouse gas inventories: Mineral industry emissions., Technical report, Intergovernmental Panel on Climate Change (IPCC)

Linton J, Walsh S (2003) From bench to business. Nat Mater 2:287–289

Mekonnen M, Hoekstra A (2016) Four billion people facing severe water scarcity. Sci Adv 2(2):e1500323

Minter A (2013) Junkyard planet: travels in the billion-dollar trash trade. Bloomsbury, New York

MIT (2013) Report of the mit taskforce on innovation and production., Technical report, Massachusetts Institute of Technology (MIT)

Mohanty DP, Udupa A, Chandra A, Viswanathan K, Mann J, Trumble K, Chandrasekar S (2021) Mechanical behavior and high formability of palm leaf materials. Adv Ener Sustain Res 2(4):2000080

Montgomery D (1997) Ecoefficiency in consumer products. Philos Trans Roy Soc A 355:1405–1414

Moynihan M, Allwood J (2014) Utilization of structural steel in buildings. Proceed Roy Soc A 470

Mufti A, Bakht B, Tadros G, Horosko A, Sparks G (2005) Are civil structural engineers risk averse? can civionics help? Sens Issues Civil Struct Health Monitor 3–12

NAE (2002) Raising public awareness of engineering. The National Academies Press, Washington, DC, US, Technical Report

Nature (2014) Editorials: Renewed energy. Nature 510:7–8

Norton R (2006) Machine design: an integrated approach. Pearson Prentice Hall, Englewood Cliffs, New Jersey, USA

Otto K, Antonsson E (1991) Trade-off strategies in engineering design. Res Eng Design 3(2):87–104

Rao B (2013) How disruptive is frugal? Technol Soc 35:65–73

Rao BC (2019) The science underlying frugal innovations should not be frugal. *Royal Society Open Science* 6(5):180421. https://doi.org/10.1098/rsos.180421

Roozenburg N, Eekels J (1995) Product design: fundamentals and methods. John Wiley & Sons, New York, USA

Rossi M, Germani M, Zamagni A (2016) Review of ecodesign methods and tools barriers and strategies for an effective implementation in industrial companies. J Clean Prod 129:361–373
Seliger G (2007) Sustainability in manufacturing. Springer, Berlin, Germany
Shanley F (1962) Historical note on the 1.5 factor of safety for aircraft structures. J Aerosp Sci 29(2):243–244
Shigley J, Mischke C (1989) Mechanical engineering design. McGraw Hill, Inc., New York., USA
Song J, Chen C, Zhu S, Zhu M, Dai J, Ray U, Li Y, Kuang Y, Li Y, Quispe N, Yao Y, Gong A, Leiste U, Bruck H, Zhu J, Vellore A, Li H, Minus M, Jia Z, Martini A, Li T, Hu L (2018) Processing bulk natural wood into a high-performance structural material. Nature 554:224–228
Stallforth P, Clardy J (2013) One size fits most. Nature 495:456–457
Steffen W, Persson Å, Deutsch L, Zalasiewicz J, Williams M, Richardson K, Crumley C, Crutzen P, Folke C, Gordon L, Molina M, Ramanathan V, Rockström J, Scheffer M, Schellnhuber H, Svedin U (2011) The anthropocene: from global change to planetary stewardship. AMBIO 40(7):739
Telenko C, O'Rourke J, Seepersad C, Webber M (2016) A compilation of design for environment guidelines. J Mech Design 138:1–11
Timoshenko S, Young D (1968) Elements of strength of materials. Van Nostrand, Princeton, New Jersey
Tollefson J (2017) Wood grows up. Nature 545:280–282
Torenbeek E (2013) Advanced aircraft design. John Wiley & Sons, New York, USA
Ugural A (2015) Mechanical design of machine components. CRC Press, New York, USA
Ullman D (2003) The mechanical design process. McGraw Hill Inc, New York, USA
US (2015) Connected vehicle pilot deployment program phase 1, concept of operations (conops), icf/wyoming. Technical Report
Valiev R, Islamgaliev R, Alexandrov I (2000) Bulk nano-structured materials from severe plastic deformation. Progr Mater Sci 45:103–189
Vezzoli C, Ceschin F, Diehl J, Kohtala C (2015) New design challenges to widely implement sustainable product-service systems. J Clean Prod 97:1–12
Vincenti W (1990) What engineers know and how they know it. John Hopkins University Press, Maryland, USA
Wise C, Pawlyn M, Braungart M (2013a) Living in a materials world: building with an eye on nature. Nature 494:172–175
Wise C, Pawlyn M, Braungart M (2013b) Living in a materials world: upcycle to eliminate waste. Nature 494:174–175
Withers P (2007) Residual stress and its role in failure. Rep Progr Phys 70:2211–2264
Wood R (2014) The challenge of manufacturing between macro and micro. Am Sci 102:124–131
Wu X, Jiang P, Chen L, Yuan F, Zhu Y (2014) Extraordinary strain hardening by gradient structure. Proceed Nat Acad Sci (PNAS) United States America 111(20):7197–7201

Chapter 7
Complex Systems-Based Design of Frugal Products

REPRODUCED FROM JOURNAL ARTICLE OF THE ROYAL SOCIETY

This chapter is derived from the author's previous work **"On complex systems of adaptive frugal products,"** ***Royal Society Open Science*, 7(7), 2020, pp. 192057, doi:** http://doi.org/10.1098/rsos.192057**,** previously published by the Royal Society under the terms of the Creative Commons Attribution License http://creativecommons.org/licenses/by/4.0/, which permits unrestricted use, provided the original author and source are credited.

7.1 Introduction

Products of the *frugal* type have attained prominence and diffused around the world (Rao 2013) (Knorringa et al. 2016) (Hossain and Halme 2016) (Pisoni et al. 2018) due to their potential to improve living standards of society at large. This is due to the affordability of quality products realized through restrictions on the consumption of resources for producing no-frills versions of electrocardiograms, cars and plasma technology, with more examples covered in the previous chapters. The examples showcased in Chaps. 3 and 4 cover *frugal* products of both the *grassroots* and *advanced* types in sectors including healthcare, automotive, astronomy and particle physics, to name a few.

Despite their popularity, such products depend critically on the design stage during which engineers and designers attempt to foresee as many issues as possible so as to realize a defect-free *advanced frugal innovation*. In this regard, the *frugal design approach* outlined in the previous chapter together with the underlying *factor of frugality* were mentioned to be executed iteratively to wean out weak spots at the initial stage of product development. However, even with a thorough design followed by a successful product, it should be noted that, generally, *frugal products* due to

© The Author(s), under exclusive license to Springer Nature Singapore Pte Ltd. 2024

B. C. Rao, *Frugal Engineering*, Design Science and Innovation,
https://doi.org/10.1007/978-981-99-9700-8_7

their usually simple and lean designs are prone to failure due to overloading. Such overloading scenarios were elaborated upon in the previous chapter. This chapter describes an interesting solution to the overloading problem that is based on the theory of complex systems. In particular, this Chapter focuses on the complex networking of products designed through the *frugal design approach* (of Chap. 6) for getting the best performance from *frugal* products. In other words, complex networking supplements the *frugal design approach* for generating top performing *frugal* products.

7.2 Background

When compared to conservative designs that employ material padding for enhancing safety (Moynihan and Allwood 2014) (Mufti et al. 2005), the no-frills nature of *frugal* products leads typically to lesser extraction of material resources and also limits emissions of greenhouse gases (GHG) (Moynihan and Allwood 2014). Therefore, low cost, lower material wastage in design and fabrication and good functionality make *frugal* products an appealing sustainable alternative to incumbents. As mentioned previously in this book, *frugality* through non-material resources, such as supply chains and sales, for lowering costs are not pursued with a focus mainly on infusing frugality at the most basic level of product development, i.e., design and engineering.

The absence of excess material padding—that is typical of conservative designs (Moynihan and Allwood 2014)—typically results in lower *factors of safety* for *frugal* products (Rao 2017b) (Rao 2019). Hence, rigorous design procedures have to be followed, as part of the *frugal design approach*, for achieving robust performance while accommodating the resulting lower *factors of safety*. The rationale for assuming that companies such as GE® and Tata Sons®, that have created the electrocardiogram and Nano Car, respectively, have undertaken rigorous design procedures to create *frugal* products is that the underlying sophistication in technology requires one.

Further, error-proofing of *frugal* devices can be achieved by connecting these products in a network through the *Internet of Things* (IoT) to enable "talk" benefitting their performance. The advent of interconnectivity between products including *frugal* ones is inevitable considering the advancement in IoT and other technologies(Schwab 2016) (NAS 2017a). The "talk" in a network would forewarn the arrival of failure conditions that could be utilized for enhancing performance or discarding affected *frugal* products, depending on the context. In fact, network "talk" would even buttress the aim of achieving low cost because a cheap no-frills structure can be complemented by information from network that would help maintain product functionality under adverse conditions.

Moreover, interconnectedness between ever-increasing numbers of *frugal* products to achieve a goal would create a complex system because of the high likelihood of failure of some *frugal* products, possibly due to perturbation from marginal overloads and, the subsequent functioning necessary of the underlying complex network through dynamic evolution of network-topology. So any attempt at making *frugal* products adaptive would greatly aid in achieving controllability of networks associated with complex systems thereby minimizing propagation of failures. In particular, adaptation would involve either product termination or morphing to strengthen for survival, in the face of adversity, thereby making a network of such products also adaptive, which by definition is complex (Grisogono 2006).

Accordingly, sections of this chapter have been organized to present the rationale underlying the complex system angle on *frugal* products and the consequent need for making individual products adaptive. Subsequently, adaptation theory with a simple example to highlight the proposed concept is presented for futuristic employment in real-time applications. Lastly, local and global networks of *frugal* products have been described to facilitate futuristic utilization with a rudimentary example involving *electric vehicles*. It should be noted that this chapter sets a framework and is therefore general in its scope since both adaptation methodology and the complex systems angle on *frugal* products have not been reported to date.

7.3 Frugal Products and Complex Systems

Frugal products are built against constrained use of resources and as such are typically vulnerable to failures under nominal conditions and/or conditions of even marginal overloading (Rao 2019). Hence a plausible solution is the adaptation against impending failures involving product fortification or termination. It should be noted that this chapter uses the term "fortify" instead of "strengthen" due to the latter's connotation of a material's yield strength or tensile strength which will be utilized in an example covered in this effort. If a variety of adaptive *frugal* products are integrated into a network to achieve a goal, there is still a significant chance of a certain proportion failing and hence such a network should function, all the time, with some failed members and network-failure happening when the proportion of failed products goes beyond a certain threshold (Cook 2000). Therefore, interconnectivity between increasing numbers of different varieties of adaptive *frugal* products, for achieving a goal, would constitute a heterogenous complex network (Malikopoulos 2015), underlying a complex system, wherein failure of some products would not alter the functionality of the network (Helbing 2013). Moreover, as a network of adaptive *frugal* products grows with time through natural evolution, as opposed to being systematically engineered, it will tend to be more complex.

The design and adaptation of *frugal* devices have a bearing on the dynamics of both individual products and their interactions and hence influence the controllability of the underlying network (Liu and Barabási 2016). In the event of adversity, adaptation would aid in retaining the functionality of at least critical *frugal* products and thereby obstruct the cascading of failure. Moreover, the affordability of complex systems in general (Kurtoglu and Tumer 2008) lends support to developing low-cost *frugal* products that can adapt against impending failures and, which subsequently are networked, for achieving a goal. In so doing, adaptation will lower, but not avoid altogether, the redundancy of products in the network. Hence it is imperative to impart adaptive abilities to individual *frugal* products for facilitating robust functioning of network under real-time conditions. Such adaptation—involving fortification or rejection of concerned *frugal* product(s)—being based on identification of weak spots through theory and their subsequent continual monitoring for impending failures through suitable sensor networks involving *Internet of Things*.

7.3.1 Linking Frugal Products

Frugal systems can span either a local or global network wherein the former refers to isolated products such as an automotive vehicle or a mobile phone and the latter denotes larger interconnectedness between local networks or products. For both local and global networks, the design and adaptation of their constituents are crucial for the effective functioning of the system as a whole and also for generating complex behavior. In fact, adaptation of individual constituents and also interactions between them will remain invariant under various perturbations and/or network topologies, especially for global networks (Gao and Barabási 2016).

In recent years, the advent of the *Internet of Things* is enabling wireless communication between a variety of devices equipped with controllers, means for wireless communication, protocols, etc., for mutual interaction and also "talk" with users in completing various tasks to fruition. In so doing, the IoT will facilitate integration of various devices with the *Internet* (Atzori and Morabito 2010), (Zanella and Zorzi 2014). Therefore, other than standard couplings, such as mechanical ones between *frugal* products, IoT should also be used, as applied in this work, to enhance connectivity and also abet product-adaptation through features of IoT enabled devices to provide for appearance of complex behavior in networks, local or global.

7.4 Methodology for Adaptation

Rigorous procedures have to be employed for designing *frugal* products due to constraints on materials consumed during their realization. In particular, lesser amounts of raw materials usually translate into lower *factors of safety*, in *classical* design, that increase the likelihood of failure arising under loading that is outside the range of safety (Rao 2019). Examples of failure due to overloading abound in various sectors such as railroad (Kim and Kang 2017) (de Faria and Santos 2017), aerospace (Infante and Baptista 2017), shipping (Valet and Beer 2013) and medical (Hernandez-Rodriguez and Garcia-Sanchez 2015), to name a few. Consequently, this chapter focuses on failures due to overloading, while risks posed by nominal conditions through material defects, product geometry and other product-based factors (Miller 2002) controlled by rigorous design in the *factor of frugality* approach of Chap. 6.

Rigorous design procedures are imperative for *frugal* products, whose exclusion can lead to failures under even nominal conditions and which might be the case for *grassroots frugal innovations* (GFIs) which are makeshift contraptions. As described in the previous chapter, even though some overloading could be sustained by a rigorously designed *frugal* product with a low *safety factor*, ~ 1.5, here an assumption is made of zero tolerance for overloading. Therefore, sensor(s) should be embedded in fortified *frugal* products around their weakest areas for continuous health monitoring so as to take remedial action when abnormal conditions, in the form of excess loading, arrive. Sensor signals from a network of fortified products will facilitate forewarning of failure on a target product that is away from the approaching failure. Consequently, two courses of action are available when conditions of failure are sensed by a *frugal* product. First, the product could be triggered to terminate functioning when there is a swarm of products collectively working to achieve a goal. Second, the weakest areas of the product can be strengthened by additional features that are instantly initiated by forewarning from sensors. In other words, sensors will abet either termination of a concerned fortified *frugal* product altogether or initiate a remedial action involving possibly morphing to withstand excess load. All of this is possible under the assumption of the availability of a window of time from receiving signal to the undertaking of actual action.

Therefore, this chapter presents a two-pronged approach for adaptation of individual *frugal* products wherein "fortifying" a product through the rigorous design of the *factor of frugality* approach is followed by "forestalling" of impending failure by *health monitoring* of critical areas through embedded sensors and subsequent remedial action. This adaptation methodology would henceforth be referred to as *fortification and forestalling* (FAF).

An existing case study on the design of a shaft (Ugural 2015) has been selected to exemplify the proposed FAF methodology of adaptation. The focus on the shaft is justified by its significance as one of the basic blocks in engineering design widely used to transmit power and/or conversion between linear and rotary motions.

7.4.1 Fortification Through Factor of Frugality

Figure 7.1 illustrates a shaft to be designed and also fabricated for *frugality* against torsional and bending loads acting on it. Gear A carries a load of 1601 N and is attached to shaft AB, made of AISI 1010 steel, whose other end is connected to gears D and E. The details of designing shaft AB by the *classical factor of safety* approach is given in Ugural (2015), where a *safety factor* (S) of 1.6 has been applied to get the results listed in Table 7.1.

The results pertaining to *factor of frugality* (F^S) for shaft AB are listed in Table 7.2, where the methodology outlined in the previous chapter has been applied to the baseline data of Table 7.1. Accordingly, the columns grouped under *Material Saved* (MS) contain scores for various extraneous material saving schemes. Each entry of Table 7.2 highlights the contribution of a scheme to conserving resources with the baseline design of Table 7.1 listed first. Here, F is equated to S since no extraneous scheme is utlized in baseline design to save material other than that resulting from a

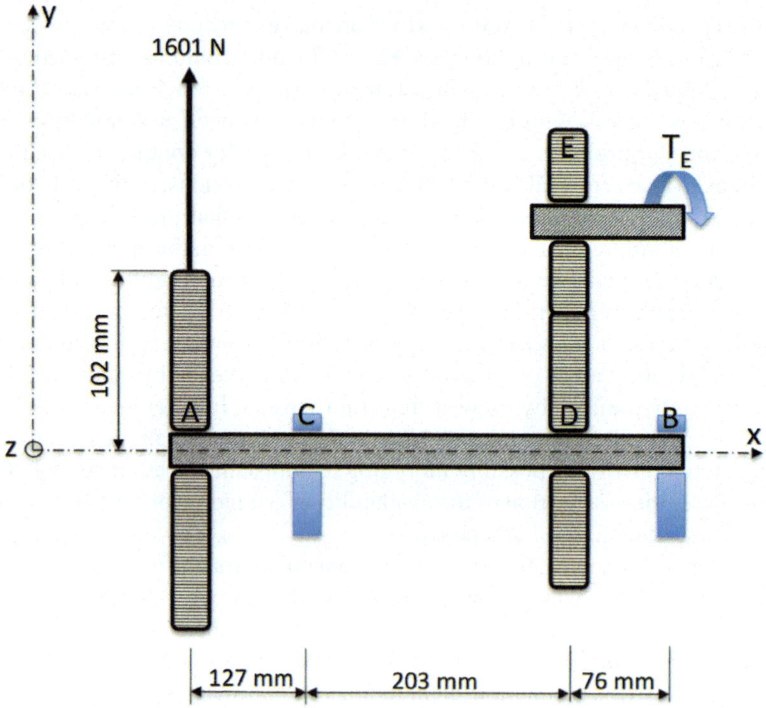

Fig. 7.1 Shaft under torsional and bending loads (Ugural 2015)

Table 7.1 Results of baseline shaft design

S. No.	Factor of safety (S)	Material	Critical section	Diameter of critical section (D mm)
1	1.6	AISI 1010CD	C	24.2

Table 7.2 *Factor of frugality* for the shaft

S. No.	Factor of safety (S)	Material saved			Factor of frugality (F)
		Simple design (MS_1)	Manufacturing (MS_2)	4R mechanisms (MS_5)	
1	1.6	0	0	0	1.6
2	1.6	0.5	0	0	2.1
3	1.6	0.5	1	0	3.1
4	1.6	0.5	1	1	4.1

design with low *factor of safety*. The remaining entries of Table 7.2 outline savings from individual extraneous material saving schemes except for the last entry, which also lists cumulative F number aggregated from S and MS values corresponding to all schemes.

The second entry of Table 7.2 quantifies material saved by replacing a solid shaft with a hollow one through MS_1 of *simple design* scheme using the maximum shear stress theory of failure (Ugural 2015). The third entry pertains to a scheme that utilizes extrusion for minimizing the wastage of excess material removed during manufacturing. Accordingly MS_2 is assigned one to reflect avoidance of having to machine the internal diameter of a solid shaft to arrive at the hollow feature. Entry 4 relates to salvaging such an extruded hollow shaft from a suitable *end of life* (EOL) system. The parameter MS_5, representing *4R mechanisms* scheme for raw material conservation through salvaging, is assigned a value of one assuming that the hollow shaft has been procured from a discarded product and not compromised in functionality. The three schemes taken together with S result in a F value of 4.1 which shows an improvement of 156 % vis-à -vis the F or S value of 1.6 for the baseline design. Therefore, the final design consists of an extruded hollow shaft, which will be procured from an EOL system, whose weakest section, shown in Fig. 7.1, is at C.

7.4.2 Forestalling by Shaft

The weakest section C should be inspected in real time by suitable structural health monitoring techniques (Farrar and Worden 2007) for forestalling failure. The shaft on notification of the arrival of excess loading could change configuration at C using morphing techniques with typical representative examples reported in (Fraunhofer-IWES 2018) and (Lachenal and Daynes 2012). Morphing, in general, would aid the development of self-healing *frugal* products (Strogatz 2001). Alternately, the shaft could shut itself, without recourse to healing, thereby deactivating several links possibly connected to it.

The FAF methodology can also be applied to other components such as and, not limited to, plates, beams, etc. It should be noted that the shaft is a representative *frugal* product and generally *frugal* devices have to be designed and adapted according to the FAF methodology prescribed here. Accordingly, an adaptive *frugal* product could be used in an isolated application or could be part of a network of only *frugal* products underlying possibly a complex system (Strogatz 2001). Although FAF could be applied to both applications, this chapter focuses on networks associated with *frugal* products since the same principles also underly an isolated *frugal* product, that is a local network of parts.

7.5 Frugal Products and Networks

This section describes two types of networks associated with *frugal* products that could materialize in the foreseeable future. Accordingly, *frugal*-product-as-a-network-of-*frugal*-components and networks-of-*frugal*-products will facilitate understanding of changes at individual product level on the system as a whole. An implicit assumption in this effort is that the price of any *frugal* product will be low due to their eventual commercial success.

7.5.1 Frugal Product as Network of Frugal Components

An individual assembly of *frugal* components is a stand-alone local network forming an isolated *frugal* product. Examples include *frugal* versions of arbitrary products including computers, mobile handsets and cars, to name a few. Although this chapter ideally supports FAF of all *frugal* components of a product due to the limited expanse of the ensuing network, forestalling in FAF can be implemented partially as shown later. Even blanket incorporation of FAF will only make *frugal* products into better adaptive systems since the bulk of current products is not designed as complex systems with suitable individual parts, their numbers and interactions commensurate with the appearance of associated features including emergence and self-organization (Barabási 2016) (Ottino 2004) (Amaral and Ottino 2004). In other words, individual *frugal* products should at a minimum be pseudo-complex systems—which are adaptive and strictly not complex—possessing *frugal* parts all of which can adapt, i.e., possess FAF capability and, possibly evolving through successive versions into proper complex systems. Therefore, currently, the bulk of *frugal* products that could be built with blanket FAF can be classified under "pseudo" due to difficulties of engineering actual complex systems at a local level.

The example of an *electric vehicle* (EV) as a *frugal* product is apt due to its ever-growing importance to *sustainable development* (UCS 2018). Although a *internal combustion engine* (ICE) vehicle could be viewed as a network of *frugal* components, an EV lends itself to networks especially due to its simplicity of structure with fewer parts and ability to network with other EVs, relevant entities and a smart grid (Shuai and Pelov 2016) (NAS 2017b). Besides, EVs are typically low-cost (UCS 2018) alternatives to ICEs, partly due to possible subsidies, which is appealing from a *frugal* viewpoint. Therefore, a *frugal* EV built out of *frugal* components could also be utilized for power generation, with a smart grid, other than its primary function serving mobility. Figure 7.2 shows a general configuration of the parts making up a *battery electric vehicle*, which is a pure EV.

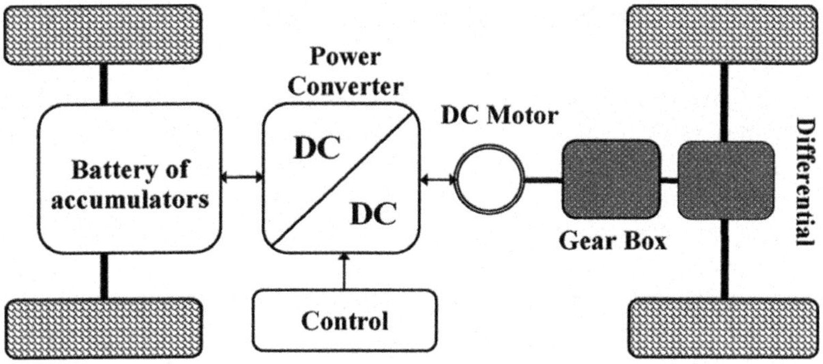

Fig. 7.2 Configuration of a battery operated or pure *electric vehicle* (Majdoub and Chaoui 2014)

7.5.2 Network of Frugal Products

As ever more *frugal* products become interconnected to each other, a global network spanning multitudes of such products would emerge. Such large-scale interconnectivity would achieve functionality requiring collections of a variety of *frugal* products, which will entail talk among products to forestall failure at the individual level including signaling backup products during failures. Although this work assumes inevitability of interconnectivity, the wide variety of *frugal* products from different sectors, such as automotive, aerospace, healthcare, etc., should at least talk to each other for sharing information/data for facilitating adaptation through FAF. Such large-scale interaction, hinting at impending dangers, in a large heterogenous network of *frugal* products possessing inherent FAF abilities coupled with external IoT connectivity features is key to achieving network complexity. Moreover, reconfiguration of a large network through continuous net addition of adaptive *frugal* products with IoT connectivity would only make the network larger (Watts and Strogatz 1998) and enhance its complexity. Successful complex networks of *frugal* products would even approach the vast expanse of the Internet (Barabási 2016) possibly in the distant future.

An instance of a large complex network would be linking of adaptive *frugal* EVs to other such EVs and other relevant entities such as grids—all of them also possessing IoT connectivity—for managing electric-power generation. It should be noted that the assumption of the entire network being *frugal* is relaxed here with only EVs being *frugal* and other entities such as grids and utilities being conventional. Such large-scale interconnectivity would result in robust functionality of both the network and its elements including *frugal* EVs. Other than managing electric power generation as a collective goal, such a global network of *frugal* EVs would also facilitate safety in mobility by nudging EVs to adapt in response to failure-related information gathered from other network entities.

7.6 More on Complexity Aiding Frugality

As shown in previous chapters, the *frugal design approach* subsumes the classical *factor of safety* while also quantifying the incorporation of other schemes that further streamline a low *S* design. This chapter outlines the use of existing *structural health monitoring* techniques in a new fortification framework to continuously monitor the weaker portion(s) of a *frugal* product designed through the *frugal design approach.* In doing so, sensor(s) are placed at locations deemed weak during design to monitor continuously and hence prevent any impending failure. Moreover, a sensor qualifying for fortification should primarily have proper functionality whose cost should be the lowest possible when compared to other suitable sensors. Although functionality takes precedence to ensure smooth operation of the *frugal* product, a low-cost sensor commensurate to proper functionality should be selected from others in consideration.

A *frugal* product is isolated and as such cannot utilize features of global complex networks such as resilience affording a product extra protection not least by giving extra time for responding to failures. Hence, a strong solution for making individual *frugal* products formidable would involve rigorous application of FAF to all their weaker sections. However, generally, such blanket application to a whole product coupled with faster reaction times should be reckoned against focusing on tempering costs. Accordingly, a weak solution would involve partial FAF for certain product topologies that could be pursued as follows. The presence of components that connect to a majority of others, i.e., hubs (Barabási 2016), would make products disassortative (Barabási 2016), where other than fortifying all product components, only hubs should be forestalled through FAF or made to have clones, i.e., backup hubs, depending on the constraint of keeping costs low. A caveat being importance of hubs assumed as solely based on their larger degree or larger numbers of connections when compared to other graphical nodes.

The architecture of a pure EV, shown in Fig. 7.2, could also be seen as a local network of *frugal* components. Accordingly, Fig. 7.3 shows *power converter* followed by *DC motor* as hubs with their significance in that sequence based on number of connections with other components or nodes. It should be noted that the shaft example given in Sect. 7.3 should be applied for the design and fabrication of axles for a *frugal* EV. Other than employing FAF for all components in Fig. 7.2, as part of a strong solution, a weaker one would involve fortification of all components but forestalling only hubs of Fig. 7.3 and excluding the remaining parts from health monitoring. The degree of cost reduction together with performance would decide the choice. The "frugalized" pure EV of Fig. 7.2 is a local or isolated pseudo-complex system whose architecture needs to be refined by considering components not shown to arrive at an accurate network representing a successful *frugal* EV.

In contrast, the interconnectedness between *frugal* products—themselves pseudo or complete complex systems—in a global complex network would provide an extra layer of protection. In particular, the inherent resilience of a global complex network would facilitate functioning in the face of localized failures by protecting *frugal*

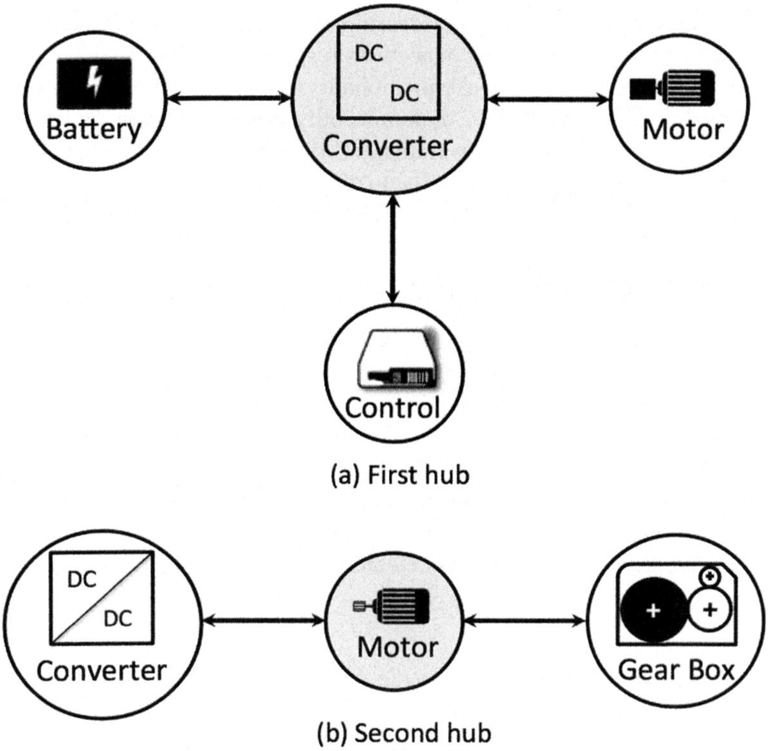

Fig. 7.3 Hubs (shown shaded) in the *electric vehicle* of Fig. 7.2

products in the network by quashing the cascading of failure through these products. Therefore, blanket FAF of individual products (local networks) together with the resilience of a global complex network would increase the resistance of individual *frugal* products against failures to a maximum, thereby facilitating robust functionality while keeping costs low.

The evolutionary growth of networks of *frugal* products is inevitable due to appealing features of both products and networks, which will be useful as a potent *sustainable* solution against the crises of our time. Therefore, non-engineered growth of a global network of *frugal* products—individually possessing inherent adaptation and network integration abilities—aiding *preferential attachment*, because of the significance of some products, would nucleate hubs thereby making the network *scale free*(Barabási 2016), a widely occurring model of small world phenomenon in complex systems (Amaral and Ottino 2004). The emergence of hubs, which are large-degree nodes, will aid accessibility of failure-prone *frugal* products of a dissortative network in a shorter duration of time which is crucial to the timely control of such networks. Even though growth is necessary for a successful *scale free* network, the negation of some *frugal* products is essential in the event of failure to prevent a cascade. The widespread morphing and severing activities in a large network of *frugal* products will strengthen or erase connections respectively, thereby making network topology commensurate with robust functionality (Gross and Sayama 2009). In other words, a *frugal* product becomes a dynamic entity whose dynamic interactions with other *frugal* products make the overall network evolve with time. Such dynamic interactions between *frugal* products due to dynamic variation of their states are also a hallmark of complex networks (Holovatch and Thurner 2017).

A possible cost-effective solution to controlling both individual products and their parent dissortative network should involve blanket fortification covering the whole network while limiting forestalling to hubs. This strategy is similar to the one applied to local networks of individual products save for application to entire products, instead of components. Consequently, the evolutionary growth of such a *scale free* network would create new hubs and make the existing ones larger (Barabási 2016), both of which will play a crucial role in controlling functioning, thereby maintaining low cost for the bulk of *frugal* products in the network. Therefore, it is imperative that networks of *frugal* products are allowed autonomous growth to be *scale free* so that corresponding complex systems have enough constituents for the emergence of hubs that relate dissortatively to other elements of their networks.

7.6.1 Network of Frugal Electric Vehicles

Electric vehicles used for power generation, besides mobility, are called *gridable electric vehicles* (GEVs) (Liu and Gao 2013), whose *vehicle to house* (V2H), *vehicle to vehicle* (V2V) and *vehicle to grid* (V2G) frameworks for connecting both home and community level grids have been reported in detail by Liu and Gao (2013). In fact, V2H, V2V and V2G systems inherently possess aggregators, which are controllers coordinating charging/discharging activities between groups of GEVs and grids. Large networks of *frugal* gridable EVs, or *frugal* GEVs, with connections to aggregators in addition to their inherent adaptability, i.e., FAF, coupled with IoT features, will show complex behavior, especially when used for power generation. As seen in Fig. 7.4, where only GEVs are assumed to be *frugal*, the presence of FAF, IoT and GEV controllers will enable the appearance of complex behavior with IoT aiding long-range talk among GEVs and other entities that could be spread across a country, which short-range V2H, V2V and V2G systems cannot penetrate. The long-range IoT wireless connectivity, enabling talk among GEVs, grids, utilities and other entities, will plausibly make its appearance through 5G networks in the foreseeable future (Akpakwu and Abu-Mahfouz 2018). Figure 7.4 is a simplistic snapshot of a hypothetical scenario that does not show dynamic features including dissapperance and apperance of links and/or nodes that bring out behavior of complex networks. Moreover, the network of Fig. 7.4 will have to be scaled up with larger numbers of both links and nodes at a country level involving a national grid system to show complex behavior.

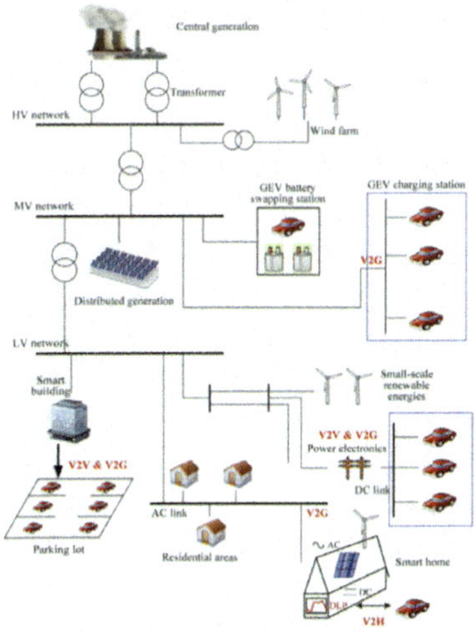

(a) Typical integration of V2H, V2V & V2G systems (Chau et al. (2013))

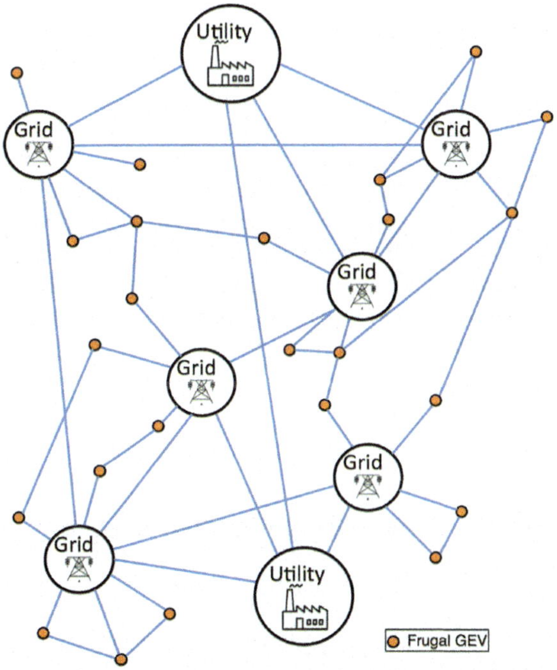

(b) Snapshot of a part of large complex network comprising frugal GEVs and relevant entities for power generation

Fig. 7.4 Complex networking of *frugal gridable electric vehicles*

7.7 Adaptation and Networking Framework for Frugal Products

The *adaptation and networking* (AAN) framework for a given *frugal* case, where adaptation refers to FAF abilities of *frugal* products comprising network, should be tested for its viability at local and global levels. The EV example shows suitability of AAN to a global network of *frugal* GEVs. Also, for a given *frugal* product, trade-offs between adaptation and networking should be considered to decide whether lesser rigor can be tolerated through networking so that initial outlays coming from rigor in both design and fabrication can be tempered. Moreover, although this effort assumes all networked products being *frugal*, the AAN framework could be extended to mixed networks of conventional and *frugal* products as in the example of a global network of GEVs. *Artificial Intelligence* (AI) techniques can be used to impart cost-effective complexity to local pseudo systems for learning from the product's environment and later using it to facilitate apt interactions between their limited parts. All in all, the use of AI could enhance feasibility and subsequently lead to fruition of coupling complex systems theory with *frugal* products to improve functionality at lower costs.

A complex network of *frugal* products will have important ramifications for *sustainable development*. The consumption of lesser resources in creating individual products would typically entail lesser numbers of activities in various parts of their *life cycles*. This would limit *carbon footprint* to the greatest extent when more and more products become *frugal*. And use of AAN would make the functionality of *frugal* products robust thereby giving them enhanced longevity and a leg-up vis-à-vis their traditional "padded" brethren. Moreover, widespread adoption of design for frugality would also facilitate effective utilization of EOL systems that could become stranded as newer green or clean technologies start getting increasingly adopted. In fact, products that are currently being streamlined for weight—to achieve lesser GHG emissions in production and operation—are *frugal* based on the terminology of this effort. Hence, multitudes of *frugal* products can make avail of AAN for giving out more, i.e., efficient functioning while taking in little, i.e., consuming lesser resources in their making and also operation.

7.8 Future: Frugal Design Complemented by Complex Networks

The advent of no-frills *frugal* products built out of constraints on resource consumption is significant for all-round *sustainable development*. This chapter has presented a *adaptation and networking* framework to offset the vulnerability of *frugal* products to failure, stemming from their structure under even marginal overloading, thereby realizing their potential of simplicity and low cost. Accordingly, complex systems theory has been invoked to account for increasing numbers of a variety of adaptive *frugal* products getting inevitably interconnected, in the foreseeable future, through *Internet of Things*-based networks. Critical to the successful implementation of large complex networks of *frugal* products is a two-pronged adaptation methodology developed here for improving product functionality. The adaptation, which is based on rigorous frugal design approach for "fortification" and health monitoring followed by remedial action for "forestalling", has been exemplified through a *frugal* shaft. Consequently, a cost-effective adaptation strategy for both local and global networks of *frugal* products is blanket fortification of all their constituents with forestalling being limited to hubs. Overall, complex behavior can be expected from larger heterogeneous networks of interconnected adaptive *frugal* products. This has been exemplified by a simple example combining gridable electric vehicles and *Internet of Things*, with individual products tending in future to local complex systems. As for *frugal* gridable electric vehicles, the current state of knowledge primarily favors a global complex network of such vehicles for collectively improving the efficacy of managing power generation/distribution with secondary spillover into the safety of mobility.

Open Problem

The example of frugal versions of *gridable electric vehicles* presented in this chapter is an interesting concept. The increasing popularity of EVs in general together with the need for energy from renewable sources makes the example on GEVs for power management suitable for actual realization. Simulate such a scenario on a suitable software/platform that simulates complex systems/networks and attempt experimentation on this concept. This is an open-ended research problem that EV companies should look into.

References

Akpakwu GA, Silva BJGH, Abu-Mahfouz A (2018) A survey on 5g networks for the internet of things: Communication technologies and challenges. IEEE Access 6:3619–3647

Amaral L, Ottino J (2004) Complex networks: Augmenting the framework for the study of complex systems. Eur Phys J B 38:147–162

Atzori LAI, Morabito G (2010) The internet of things: A survey. Comput Networks 54:2787–2805

Barabási A (2016) Network science. Cambridge University Press

Cook RI (2000) How complex systems fail. University of Chicago, Cognitive technologies Laboratory

de Faria GL, Godefroid LBLC, Santos D (2017) Failure analysis of a sleeper anchor model used in railroads. Eng Failure Anal 79:565–574

Farrar CR, Worden K (2007) An introduction to structural health monitoring. Philos Trans Roy Soc A 365:303–315

Fraunhofer-IWES (2018) Smartblades2 project: Field measurement campaign on wind turbine in colorado begins. Fraunhofer Institute for Wind Energy Systems

Gao JBB, Barabási A (2016) Universal resilience patterns in complex networks. Nature 530:307–312
Grisogono A (2006) The implications of complex adaptive systems theory for c2. Technical report, Defence Science and Technology Organisation (DSTO), Australia
Gross T, Sayama H (2009) Adaptive networks: theory, models and applications (Understanding complex systems). NECSI and Springer, USA and Germany
Helbing D (2013) Globally networked risks and how to respond. Nature 497:51–59
Hernandez-Rodriguez MAL, Contreras-Hernandez GRAJ-HBB-R, Garcia-Sanchez E (2015) Failure analysis in a dental implant. Eng Failure Anal 57:236–242
Holovatch YRK, Thurner S (2017) Complex systems: physics beyond physics. Eur J Phys 38:023002
Hossain MHS, Halme M (2016) Can frugal go global? diffusion patterns of frugal innovations. Technol Soc 46:132–139
Infante V, Fernandes LMF, Baptista R (2017) Failure analysis of a nose landing gear fork. Eng Failure Anal 82:554–565
Kim JH, Jin JWJL, Kang K (2017) Failure analysis for vibration-based energy harvester utilized in high-speed railroad vehicle. Eng Failure Anal 73:85–96
Knorringa P, Pesa I, Leliveld A, van Beers C (2016) Frugal innovation and development: aides or adversaries? Eur J Develop Res 28(2):143–153
Kurtoglu T, Tumer I (2008) A graph-based fault identification and propagation framework for functional design of complex systems. ASME J Mech Design 130:051401
Lachenal XPW, Daynes S (2012) Multi-stable composite twisting structure for morphing applications. Philos Trans Roy Soc A 468:1230–1251
Liu Y, Barabási A (2016) Universal resilience patterns in complex networks. Rev Modern Phys 88:035006
Liu C, Chau KTDW, Gao S (2013) Opportunities and challenges of vehicle-to-home, vehicle-to-vehicle, and vehicle-to-grid technologies. Proceed IEEE 101(11):2409–2427
Majdoub KEI, Giri FHO, Chaoui FZ (2014) Nonlinear cascade strategy for longitudinal control of electric vehicle. ASME: J Dyn Syst Meas Control 136(1)
Malikopoulos A (2015) Centralized stochastic optimal control of complex systems. Eur Control Conf (ECC) 15–17. Linz, Austria 978-3-9524269-3-7 2015:721–726
Miller B (2002) ASM handbook: failure analysis and prevention: overload failures. ASM International, USA
Moynihan M, Allwood J (2014) Utilization of structural steel in buildings. Proceed Roy Soc A 470
Mufti A, Bakht B, Tadros G, Horosko A, Sparks G (2005) Are civil structural engineers risk averse? can civionics help? Sens Issues Civil Struct Health Monitor 3–12
NAS (2017a) The fourth industrial revolution: Proceedings of a workshop, National Academies Press (US), Washington (DC)
NAS (2017b) Strategies to advance automated and connected vehicles. Technical Report, National Academies of Sciences, Engineering, and Medicine, Washington, DC, USA
Ottino JM (2004) Engineering complex systems. Nature 427:399
Pisoni A, Michelini L, Martignoni G (2018) Frugal approach to innovation: State of the art and future perspectives. J Clean Prod 171:107–126
Rao B (2017a) Advances in science and technology through frugality. IEEE Eng Manage Rev 45(1):32–38
Rao B (2017b) Revisiting classical design in engineering from a perspective of frugality. Heliyon 3(5):e00299
Rao B (2019) The science underlying frugal innovations should not be frugal. Roy Soc Open Sci 6(5)
Rao B (2013) How disruptive is frugal? Technol Soc 35:65–73
Schwab K (2016) The fourth industrial revolution. World Economic Forum
Shuai WPM, Pelov A (2016) Charging electric vehicles in the smart city: A survey of economy-driven approaches. IEEE Trans Intell Transp Syst 17(08):2089–2106
Strogatz S (2001) Exploring complex networks. Nature 410:268–276

UCS (2018) Cleaner cars from cradle to grave: How electric cars beat gasoline cars on lifetime global warming emissions. Technical report, Union of Concerned Scientists, USA
Ugural A (2015) Mechanical design of machine components. CRC Press, New York, USA
Valet S, Piskoty GSMCA, Beer M (2013) Accident caused by dynamic overloading of a ship mooring rope. Eng Failure Anal 35:439–453
Watts D, Strogatz S (1998) Collective dynamics of small-world networks. Nature 393:440–442
Zanella A, Bui NACLV, Zorzi M (2014) Internet of things for smart cities. IEEE IOT J 1(1):22–32

Chapter 8
Frugal Manufacturing

REPRODUCED FROM JOURNAL ARTICLE OF THE ROYAL SOCIETY
This chapter is derived from the author's previous work **"Frugal manufacturing in smart factories for widespread sustainable development,"** ***Royal Society Open Science*, 8, 2021, pp. 210375, doi:** http://doi.org/10.1098/rsos.210375, previously published by the Royal Society under the terms of the Creative Commons Attribution License http://creativecommons.org/licenses/by/4.0/, which permits unrestricted use, provided the original author and source are credited.

8.1 Introduction

Manufacturing is a crucial activity of product development that feeds into and is also influenced by the design process. Any material conservation gained during manufacturing directly affects the green credentials of a product. Manufacturing waste can be contrived to approach zero through the *frugal design approach* that quantifies resource conservation at all stages of development of a product engineered for frugality. Accordingly, this chapter presents *frugal manufacturing*, integral to the *frugal design approach*, for utmost reduction of waste while aiming for good surface integrity, better properties, minimal number of processes and low cost. Other than saving on energy and hence emissions, the new concept of *frugal manufacturing* also goes beyond current *near net shape* technologies, which advocate mainly for zero wastage and suitable properties while using a narrow range of manufacturing processes. Case studies involving *high speed machining*, *superplastic forming* and *additive manufacturing* of aerospace alloys have been presented that bring out the features and benefits of *frugal manufacturing*. As such the multipronged objectives of *frugal manufacturing* should be dovetailed with those of *smart factories* for creating novel technologies that abet widespread sustainable development. Such enhancement of the *smart factories* concept has been argued in this chapter to support unusual applications such as the fight against pandemics including the current one involving COVID-19.

© The Author(s), under exclusive license to Springer Nature Singapore Pte Ltd. 2024

B. C. Rao, *Frugal Engineering*, Design Science and Innovation,
https://doi.org/10.1007/978-981-99-9700-8_8

Frugality in resource consumption is an important pillar of *sustainable development* (Rao 2013; Hoekstra and Wiedmann 2014). This has been exemplified, in recent years, by the proliferation of both grassroots and sophisticated or *advanced frugal* products, which provide good (or even better) functionality, vis-à-vis existing conventional types, at low cost while consuming minimal resources (Zeschky et al. 2011; Tiwari and Herstatt 2012; Knorringa et al. 2016; Prabhu 2017; Rao 2017a; Miesler et al. 2020). Examples of low-cost electrocardiograms (Rao 2013), portable ultrasound (Rao 2013), lighting without power source (Rao 2017a) and, cars (Rao 2013), among other examples (Rao 2013, 2017a), attest to the potential of *frugal* products for widespread *sustainable development*. The *frugal* nature of these products fits in well with the cradle-to-cradle concept of sustainability promoted by the *Environmental Protection Agency* (EPA) (EPA 2015) and the *National Academy of Engineering* (NSTC 2013; Nature 2014) to make manufacturing strong and competitive (NAS 2019). Although systematic design (Rao 2017b, 2019) and subsequent complex-networking (Rao 2020) of products together with suitable policymaking and educational programs for their diffusion (Liefner et al. 2020) are crucial, manufacturing aspects are equally important for their successful realization. This is because fabrication of these products needs judicious use of manufacturing and/or assembly for lowering consumption of raw materials. Moreover, manufacturing itself plays a significant role in providing crucial inputs to the *frugal design approach* (Rao 2017b). Therefore, there is a need to make manufacturing *frugal*.

It should be noted that this effort differentiates between *near net shapes* (NNSs) (Marini et al. 2017) and *net finished shapes* (NFSs) with the former needing some form of finishing operation(s) to arrive at a NFS. In other words, an NFS is the final product ready for application.

8.2 Frugal Manufacturing

This chapter defines *frugal manufacturing* (FM) as fabrication utilizing a minimum number of low-cost processes producing zero waste for creating a NFS possessing necessary geometrical tolerances; requisite *surface integrity*; and appropriate properties. *Surface integrity* refers to the finish and accompanying sub-surface residual stresses of the product and properties refers to any additional requirements such as mechanical ones of strength, toughness, etc or any characteristic required for functionality. In other words, *net finished shapes* of requisite quality are to be produced at low cost with ideally a single process using enough quantity of material for avoiding any extra operations, including finishing and, their resulting waste. In fact, cost reduction will be further buttressed through both avoidance of waste due to concomitant need for lesser stock of feed material and a single process that negates the outlays over multiple ones. The restraint on number of processes would also lower power consumption and the attendant emissions of *green house gases* (GHGs).

Overall, FM subsumes the definition for an ideal NNS given by Altan and Miller (1990). Some efforts on NNS document both reduction of waste in specific manufacturing processes and achievement of specific properties with low defect rates. This is borne out by few existing papers (Kudo 1990; Marini and Corney 2017; Marini et al. 2017) that report aiming for a *near net shape* while also focusing on attendant functionality, mechanical properties and microstructure without explicit reference to *surface integrity*. *Surface integrity* considerations are significant because of their potential to enhance product life (Withers 2007; Lewandowski and Seifi 2016), which is vital to *sustainable development*. Also, NNS processes are typically *bulk* and *sheet* deformation operations (Cominotti and Gentili 2008) involving large plastic strains whereas *frugal manufacturing* includes the entire gamut of additive and subtractive processes. Therefore, *frugal manufacturing* is a general concept that also aims for these additions not available in NNS processes: proper *surface integrity* of the product; consideration of a wider range of processes; minimal number of processes; and low-cost process(es).

The increasing numbers of *advanced frugal* products (Rao 2017a) make it imperative to use minimal manufacturing for creating *net finished shapes*, including intricate geometries, without secondary finishing operations such as machining and grinding. This is because manufacturing is a critical input to the *frugal design* of such products and hence needs to generate as little waste as possible to qualify for *frugality* (Rao 2017b, 2019). In particular, ideally, a *net finished shape* with zero wastage would help leading to maximum improvement in the *factor of frugality* of a product (Rao 2017b, 2019). It is noteworthy that even without a *frugal* mindset, the premise underlying FM will go a long way in supporting *sustainability* through lesser wastage; lower costs due to cost-effective processes and also attendant savings in materials wasted; lower power consumption; and low-cost quality products. In fact savings in resources are especially important when processing expensive materials like superalloys (Marini et al. 2017).

8.2.1 Why Frugality in Smart Factories?

The potential of *frugal manufacturing* to achieve high quality, zero waste and, low cost in a minimal number of operations should be harnessed for the development of *smart factories* which are vital to Industry 4.0 (Liu and Xu 2017). The lesser numbers of operations—ideally a single operation—also makes FM an energy saver, which is critical to Industry 4.0. Inclusion of FM in Industry 4.0 is essential since FM controls actual operations based on their relevant physics, which Industry 4.0 does not. But realization of FM in real-time *smart factories* will require, inter alia, interactions between machine-tool systems and existing tool-kit of Industry 4.0 including digital twins and *artificial intelligence* (AI), to name a few. In fact, building of novel machine tools, designed from the perspective of FM, would be greatly aided by these pillars of Industry 4.0. All in all, incorporating FM into Industry 4.0 will make factories truly "smart" by "frugalizing" their process(es) against climate change.

8.3 Candidate Processes for Frugal Manufacturing

This section covers primary ways of manufacturing for metals that qualify for *frugal manufacturing*. Any other real-time process could be viewed as a variant of these operations—some of which are also described in this section. Although these processes have been listed separately to show their individual potential for going *frugal*, their combinations, as explained later, also aid frugality. In other words, these processes are the current building blocks of FM with possible foreseeable revisions.

8.3.1 Casting

Casting is a popular *classical* manufacturing process (Tateno 1985) that involves production of *near net shapes* through pouring; subsequent filling of mold and resulting solidification of parts. Cast parts typically need post treatments including heat and/or finish machining for imparting proper *surface integrity* and also necessary properties to the NNS. Variants to casting include NNS processes such as *pressure infiltration casting* (Cook and Werner 1991) and *gel casting* of metallic powders (Gilissen et al. 2000) while *semi solid casting* can create *net finished shapes* (Kang et al. 2003).

8.3.2 Bulk Deformation and Sheet Processing

These operations utilize severe plastic flow to arrive at parts both plain and intricate. Examples, to name a few, include extrusion, forming, rolling and forging for *bulk* deformation processes and blanking and shearing for *sheet- metal* operations. The *bulk* and *sheet-metal* processes involve minimal to no change in the volume of feed material used for fabrication. Since volumetric change is absent in plastic deformation, volume here refers only to that of the feed material. Moreover, production of intricate patterns and shapes is possible (Morita et al. 1991) through hot working and feed materials that show *superplastic* behavior as in *semi solid forming* (Witulski et al. 1994; Wilde 2014). The work by Tateno (1985) attests to the capability of these processes in fabricating large NNSs. Since these processes hew very close to the finished-part with minimal attendant finishing operations, they possess the ability to directly produce a *net finished shape* (Cominotti and Gentili 2008). In fact, roll forming can directly create NFSs for even complex geometries (Bewlay et al. 2003). However, since the quantum of feed stock going into any *bulk* or *sheet* operation is limited, the stock-quality in terms of material-defects, homogenity in composition, surface texture and other features should be superior for creating defect-free products (Kudo 1990).

8.3.3 Additive Manufacturing (AM)

The *additive manufacturing* (AM) of metallic materials has come a long way to create products with complex shapes; repair worn parts and generate parts with composition-gradient (Herderick 2011). In so doing, AM is also not ridden with problems such as machine-tool chatter that plague conventional or subtractive processes (Merritt 1965). AM or *3D printing* uses layer-by-layer deposition to build a part from ground up. Some AM techniques such as *direct metal deposition* (DMD) can be looked upon as a variant of the *casting* process. The resulting parts hew closely to the required shape but typically possess inadequate surface finish; improper residual stresses; and also possibly anisotropy in properties. Hence, AM processes typically produce NNSs (Herderick 2011) needing subsequent heat treatments and/or finishing operations.

8.3.4 Powder Metallurgy (PM)

Metallic powders are molded by *powder metallurgy* (PM) techniques into green compacts and subsequently sintered into *near net shapes* including intricate geometries. *Powder metallurgy* techniques are employed in several sectors to create products with requisite properties. However, porosity is an ever present problem (Yamamoto et al. 2010) in PM products that hampers both bulk-properties and *surface intergrity*. A majority of *powder metallurgy* techniques produce NNSs needing subsequent heat treatments, not least for minimizing porosity and, also finishing operations (Bassini et al. 2019).

8.3.5 Non Traditional Techniques

Non-traditional manufacturing (NTM) applies non-traditional mediums such as water for processing materials into products (Benedict 1987). An example of NTM is *abrasive water jet* (AJW) cutting which utilizes a water-based abrasive-slurry aimed under very high pressures to cut materials and hence make parts. These techniques typically remove very little material and hence are potential candidates for *frugal manufacturing*. However, NTM also results in NNSs that need post treatments and secondary finishing operations for achieving suitable properties and proper *surface integrity*. Besides, NTM processes are also energy intensive.

8.3.6 Metal Cutting

Metal cutting operations create parts by subtraction of material and are either of the machining or abrasive-cutting type. The former uses a cutter with well-defined geometry while the latter uses random grains, such as those of sand, bonded together in a wheel for cutting materials. In either type, excess material is removed to make a product against constraints of *surface integrity* and dimensional tolerances. *Metal cutting* typically follows the *frugal*-processes of this section as a finishing operation. However, in recent years, *high speed machining* (HSM) has shown the ability to manufacture features in a single pass, an example of which has been subsequently presented in this work as a case study. Moreover, *metal cutting* operations are also being optimized for hewing to product-shapes and hence remove least amount of stock for achieving requisite tolerances (Park et al. 2009), all akin to production of NNSs. Also, recently, *metal cutting* has been unusually utilized as a single pass *severe plastic deformation* (SPD) process for producing novel *ultra fine grained* (UFG) materials (Saldana et al. 2010) whose lesser usage, due to their superior mechanical properties, create *frugal* products.

8.4 Gauging Frugality of Processes

Frugal manufacturing's goal of restricting waste and maintaining both appropriate properties and good *surface integrity* while costing less can be realized by the individual processes outlined in the previous section or their combinations. In this regard, an existing scheme based on the *factor of frugality* can be used to quantify the streamlining of manufacturing processes going into the making of a product, *frugal* or otherwise.

8.4.1 Factor of Frugality

The *factor of frugality* (F), underlying the *frugal design approach* (FDA), facilitates systematic design of *frugal* products (Rao 2017b, 2019). The *factor of frugality* is an extension of the *classical* and popular *factor of safety* (S) to account for restraints on resource consumption, which is vital to *sustainable development*. It is the sum total of the *factor of safety* and *material saved* (MS) parameters, which account for savings in addition to that realizable from a low *S* design. In other words, *frugality* is achieved in a systematic manner by starting with a low *S*, ~1.5, design and subsequently selecting extraneous materials-and-design related features to save yet more material for this design. Both *F* and *S* are symbolically summarized in F^S with the defining equation for *F* being:

$$F = S + \sum_{i=1}^{n} MS_{\mathrm{i}} \tag{8.1}$$

Here i indexes through various extraneous schemes for conserving material including simple design, biomimetics, materials, manufacturing and salvaging (Rao 2017b, 2019). The MS parameter associated with manufacturing, i.e., FM, is denoted by MS_{MANU}. In particular, MS_{MANU} quantifies material wasted in manufacturing, e.g., as chips in *metal cutting* operations and, hence is a crucial parameter underlying FM. Accordingly, the defining equation for MS_{MANU} is given by,

$$MS_{MANU} = \frac{(W_{\mathrm{MAX}} - W_{\mathrm{MIN}})}{W_{\mathrm{MAX}}} \tag{8.2}$$

where, W_{MAX} refers to actual manufacturing operation(s) maxing out on waste production while W_{MIN} refers to operation(s) reducing waste to its lowest level and therefore selected for *frugal manufacturing*. Therefore, *frugal manufacturing's* aim of minimizing waste is achieved through maximization of MS_{MANU}. However, maximization of MS_{MANU} is sought in this work against constraints on costs and product-quality in terms of tolerances, properties and *surface integrity*. In other words, MS_{MANU} should ideally be unity by accounting for a single-pass process that is low cost and imparts necessary properties and also *surface integrity* to the part. This would also account for the need to enshrine *surface integrity* requirements in the *factor of frugality* as expounded by Rao (2017b). In other words, the model for F^{S} should also stress on achieving the correct *surface integrity* in addition to properties by focusing on MS_{MANU}.

8.4.1.1 Example on Factor of Frugality

The computation of F^{S} is exemplified by a *frugal shaft* from Rao (2019). The example is repeated here to help the reader remember the concepts and steps of the *frugal design approach*. It should be remembered that the simple example of a shaft—a popular engineering component—is useful in understanding the quantification of relevant *MS* parameters and subsequently F^{S}. The shaft, made of steel, has to be designed for transmitting a power of 500 kW at 1200 RPM under torsional loads. The details of the classical design process have been outlined in Ugural (2015). Accordingly, Table 8.1 and Fig. 8.1 present the *frugal* design of the shaft where MS_2, i.e., manufacturing's share MS_{MANU}, is just one of the *material saved* parameters going into the computation of F^{S}.

The first step involves fixing the *factor of safety* at a low admissible value of 1.5 throughout quantification. A *S* of 1.5 by itself creates a "leaner" shaft that avoids excess material accruing with increasing *S*. The next step listed as first entry in Table 8.1 concerns baseline design, i.e., classical design devoid of frugality, of a

Table 8.1 Computing *factor of frugality* for a shaft (Rao 2019). (material: steel; yield shear strength: 300 MPa; shear modulus: 80 GPa)

S. no.	Factor of safety (S)	Material saved			Factor of frugality (F)
		Alternative design (MS_1)	Manufacturing (MS_2)	Salvaging (MS_3)	
1	1.5	0	0	0	1.5
2	1.5	0.5	0	0	2.0
3	1.5	0.5	1	0	3.0
4	1.5	0.5	1	1	4.0

solid or "bulky" shaft for the S of 1.5. This results in a solid shaft of diameter 47 mm. The MS schemes are next applied with second entry, on the *material saving* scheme of *alternative design*, listing material-savings of 50% (or an MS_1 of 0.5) by designing a hollow shaft, which is the lightest candidate in the pool of possible designs for the shaft. The resulting outer and inner diameters of the hollow shaft are 56 mm and 44 mm respectively. Subsequent steps/entries consider MS schemes on *manufacturing* and *salvaging* in this respective sequence. Manufacturing is accomplished by extrusion as an NFS process entailing zero wastage and hence a MS_2 (or MS_{MANU}) of 1. The last scheme on *salvaging* pertains to procurement of the extruded-hollow-shaft from a discarded or *end of life* (EOL) system, assuming it is available. Hence MS_3 is also 1 because of complete replacement with a salvaged shaft. Both Table 8.1 and Fig. 8.1 follow the natural sequence in the development of a product such as a shaft. The final value of F, as per Eq. (8.1), is the sum of S and all the MS values. Consequently, the final F of 4, for the frugally designed shaft, is an improvement of 167% over baseline value of 1.5, due to both low S and various MS values. The rigor in applying design principles, due to a low S, though not apparent for this simple example is very significant for the success of the *frugal design approach*.

There is progressive savings in steel and, hence cost and energy, while sequencing through the steps of F^S computation. The low S gives a leaner shaft with lesser steel, vis-à-vis a conservative design, that lowers costs and also energy consumed. For a fixed price of steel and also inexpensive manufacturing, the costs reduce in each MS scheme by nearly the same proportion of steel saved. So overall there are significant savings in steel consumption and hence corresponding reductions in costs for the total F of 4. As for energy, assuming reasoning similar to reductions in costs results in similar savings. Therefore, this simple example attests to the large savings in materials and hence large reductions in both emissions and costs resulting from adoption of the *frugal design approach*. Moreover, Eq. (8.1) is a first generation model of F^S whose evolution in future would account for other *material saving* schemes and also improve its use in design.

Other than the shaft-example on computation of F^S, quantification of just MS_{MANU} for *metal cutting* is also presented, in the section on case studies, to understand its working in real time. *Metal cutting* has been selected because it is an inherently

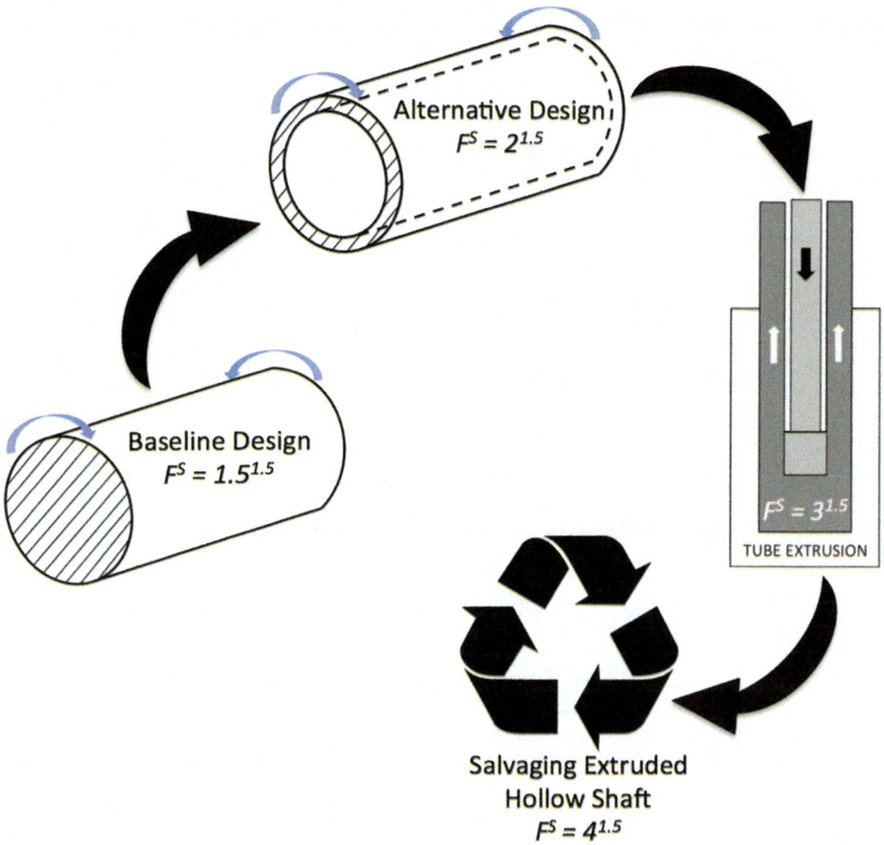

Fig. 8.1 Improvement in *factor of frugality* of a shaft through various *material saving* schemes, including manufacturing, considered in a natural sequence (Rao 2019)

wasteful operation in terms of debris generated when compared to other processes. After this brief digression into computing F^S, the remainder of this chapter will continue with the manufacturing aspect (related to MS_{MANU}) of *frugal* design.

8.4.2 *Rating Manufacturing Process(es)*

Table 8.2 highlights potential use of candidate processes of Sect. 8.3 as stand alone FM operations. Other than zero waste, each of these processes could plausibly produce a NFS at low cost in a single step with requisite microstructure, properties, *surface integrity* and geometrical tolerances. Therefore, Table 8.2 lists maximum MS_{MANU} values, by strictly adhering to the principles of FM, for individual processes independent of any application.

Table 8.2 Ideal potential for stand alone *frugal manufacturing* in individual processes

No.	Manufacturing process	MS_{MANU}	Output
1	 **Casting**	1	Net finished shape
2	 **Additive Manufacturing (AM)**	1	Net finished shape
3	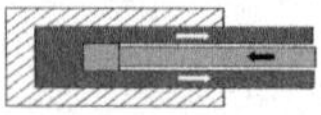 **Bulk Deformation & Sheet Processing**	1	Net finished shape
4	 **Non Traditional Manufacturing**	1	Net finished shape

However, each of the processes of Table 8.2 and, Powder Metallurgy (PM), individually leave a finish with accompanying residual stresses and/or properties, any or all of which need improvement during actual production. Table 8.3 therefore covers combinations of processes that could be utilized in real time for realizing a NFS through FM. Therefore, the primary processes of Table 8.2 produce a NNS, under existing technology, which is subsequently transformed to an NFS through secondary operations. Consequently, *MS* falls between a range of values for these process-combinations while realizing a NFS since material will have to be removed for achieving proper properties and also *surface integrity*. Other than metal removal, production of such NFSs would also entail use of surface and heat treatments with the associated cost increases for going *frugal*. Tabulations such as Tables 8.2 and 8.3 should facilitate planning and implementation of any real-time FM activity.

Table 8.3 Real time *frugal manufacturing* through combinations of processes

No.	Combinations of manufacturing processes	MS_{MANU}	Output
1	 **Casting & Metal Cutting**	$0 < MS_{MANU} < 1$	Net finished shape
2	 **Additive Manufacturing (AM) & Metal Cutting**	$0 < MS_{MANU} < 1$	Net finished shape
3	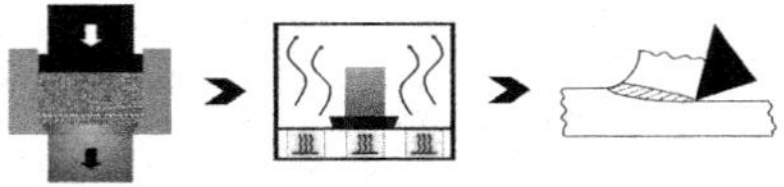 **Powder Metallurgy (PM) & Heat Treatment & Metal Cutting**	$0 < MS_{MANU} < 1$	Net finished shape
4	 **Bulk Deformation & Metal Cutting**	$0 < MS_{MANU} < 1$	Net finished shape
5	 **Non Traditional Manufaturing & Metal Cutting**	$0 < MS_{MANU} < 1$	Net finished shape

Table 8.4 Manufacturing processes from existing studies that could be "frugalized"

No.	Existing NNS process	Tolerances	Surface integrity	Properties	Process cost	Number of processes	Wastage
1	Centrifugal casting and machining (Marini and Corney 2017)	Yes	No	Yes	Unknown	3	0
2	Roll forming (Bewlay et al. 2003)	Yes	No	Yes	Low	1	0
3	Isothermal forging (Morita et al. 1991)	Yes	No	Yes	Unknown	1	Minimal
4	Flow forming (Cominotti and Gentili 2008)	Yes	No	Yes	High	1	0
5	Cold forging of spur gear (Kudo 1990)	Yes	No	Yes	Unknown	2	Minimal
6	Cold forging of tulip shaft (Kudo 1990)	Yes	No	Yes	Unknown	3	Minimal
7	Cold extrusion of spline shaft (Onodera and Sawai 1992)	Yes	No	Yes	Unknown	4	0
8	Hot forging of four stroke crank shaft (Yamagata 2005)	Yes	No	Yes	Low	7	Minimal
9	Superplastic forming of turbine disc (Moore et al. 1970; Sherby and Wadsworth 1985)	Yes	No	Yes	High	4	0

8.4.3 *Frugality from Near Net Shape Processes*

After delving into individual and combinations of operations for FM, this section presents some existing instances. Table 8.4 lists past efforts on *near net shapes* used in this study to make the case for FM since these possess several requisite features. It should be noted that some of this information has been inferred where not explicitly mentioned. Overall, these NNS efforts generate parts with requisite tolerances and also proper properties but most of them lack other features of FM. As seen in Table 8.4, none of the entries focus on *surface integrity*, which is crucial to enhancing product life and, hence *sustainable development*. The control of *surface integrity* is important, not least for avoiding fatigue (Withers 2007; Lewandowski and Seifi 2016), a predominant mode of failure arising from surface flaws. This lack of focus on the quality of both surface and sub-surface regions is present even for zero wastage, which along with tolerances and properties are touted as major benefits of NNS techniques. Furthermore, a majority of these NNS entries are not single process with several of them being expensive.

8.5 Case Studies on Frugal Manufacturing

Three different case studies on popular manufacturing operations have been selected for demonstrating the efficacy of *frugal manufacturing*. Accordingly, Table 8.5 lists details about these operations under relevant columns. The examples on NNS operations, i.e., superplastic forming and additive manufacturing, are presented qualitatively with quantification of frugality exemplified by focusing for clarity on machining due to its intrinsic higher wastage. A *near net shape* technology can be made *frugal* since it already possesses some features necessary for the transition. In particular, a minimal stock of feed material for producing parts with requisite shape, geometry and some properties is a step in the right direction. In this regard, entry 9 of Table 8.4 on superplastic forming of a nickel based superalloy, IN100, has been selected to illustrate the first case for FM. Consequently, the first entry in Table 8.5 shows the original set of NNS processes (Moore et al. 1970; Sherby and Wadsworth 1985) and also modifications for making it *frugal*. As seen in this table and Fig. 8.2, the dies used for superplastic forming could be readily retrofitted or made anew with built-in mechanism for selective heating so as to complete the heat treatment toward the end of superplastic forming. Since the die and chamber walls already possess mechanisms to heat them to superplastic temperatures, the retrofitting to control selective portions of die for post-treatment is very much possible. It should be noted that Fig. 8.2 is a general schematic of superplastic forming showing the retrofitting necessary for going *frugal* and is not a sketch of the example pertaining to turbine blades. Furthermore, the built-in selective-heating arrangement of the *frugal* process can also impart necessary residual stresses, through heat- and associated quenching-treatment (Zhang et al. 2019a), to sub-surface layers of the turbine. In so doing, FM does away with the post treatments thereby minimizing total number of processes and hence avoiding wastage of associated resources while lowering costs. In particular, when compared to separate heat treatments, cost will be reduced due to the need for only controllers that selectively heat portions of die-body in the setup. Implicit in this argument is the realistic assumption of selecting effective and also low-cost controllers. It is noteworthy that the overall heating facilities available in the setup could be used to combine even the *hipping* and warm-pressing operations of the conventional-process, thereby leading to a single-pass arrangement.

Other than NNS technology, an important technique with significant potential to go *frugal* is *additive manufacturing*. Accordingly the second case study focuses on *selective laser melting* (SLM) of Ti-6Al-4V (Zhang and Attar 2016), a popular titanium superalloy, for depositing biomedical products. Table 8.5 lists the details of this AM process in entry #2, where a single pass of SLM deposits these parts. However, the resulting parts are columnar-grained possessing acicular α' and β phases as opposed to the $\alpha + \beta$ phase combination characterizing a proper Ti-6Al-4V alloy, see Fig. 8.3 (Zhang and Attar 2016). Thus not only do SLM-deposited Ti-6Al-4V parts comprise invalid phases, thereby leading to a different material, their grain structure is columnar, which generally leads to severe anisotropy in properties (Chandra and Rao 2017). Moreover, the invalid martensite-based acicular α' phase also makes

Table 8.5 Case studies of processes going *frugal*. For clarity cost reductions have been described in the main text

Case study	Manufacturing process(es)		Properties		Surface integrity		Wastage	
	Conventional	Frugal	Conventional	Frugal	Conventional	Frugal	Conventional	Frugal
Superplastic forming of IN100-superalloy turbine disk with blades (Moore et al. 1970) (Sherby and Wadsworth 1985)	Hot isostatic pressing (*hipping*) to get billet	Same as NNS	Coarse and fine grains in blades and disk for creep- and fatigue-strength respectively	Coarse and fine grains in blades and disk for creep- and fatigue-strength respectively	Does not generate proper residual stresses	Can generate proper residual stresses through selective heating mechanism and quenching	Minimal (see Tables 8.2 and 8.3)	Same as NNS
	Warm pressing	Same as NNS						
	Superplatic forming	Combined superplatic forming and selective heating of die-portions to treat blades						
	Selective heat treatment of blades							

(continued)

Table 8.5 (continued)

Case study	Manufacturing process(es)		Properties		Surface integrity		Wastage	
	Conventional	Frugal	Conventional	Frugal	Conventional	Frugal	Conventional	Frugal
Selective laser melting (SLM) of Ti-6Al-4V-superalloy for biomedical applications (Zhang and Attar 2016)	SLM to deposit Ti-6Al-4V part	Deposit Ti-6Al-4V alloy along with inocoulant on a preheated substrate	Columnar-grain structure with atypical phases of $\alpha' + \beta$ in SLM deposition (Generates different alloy with anisotropy) Columanr grains are subsequently transformed to equiaxed type through heat treatment and/or plastic deformation	Generates proper alloy with $\alpha + \beta$ phase combination in a single SLM pass possessing necessary equiaxed microstructure and hence proper mechanical properties	Generates proper residual stresses after heat treatment and/or plastic deformation	Generates proper residual stresses in a single pass	Minimal (see Tables 8.2 and 8.3)	Minimal (see Tables 8.2 and 8.3) (Will be zero with futuristic advances in AM)
	Heat treatment and/or plastic deformation	Light finish machining to get requisite *surface texture*						
	Finish machining to get requisite *surface texture*							

(continued)

Table 8.5 (continued)

Case study	Manufacturing process(es)		Properties		Surface integrity		Wastage	
	Conventional	Frugal	Conventional	Frugal	Conventional	Frugal	Conventional	Frugal
Machining of Al7075-alloy to generate planar features for aerospace applications (Rao and Shin 2001)	Rough machining for maximum material removal. Depth of cut: 6 mm	Single cut to generate requisite finish and also compressive residual stresses (Cutting speed: 1350 m/min Low depth of cut: 0.7 mm) Needs lesser stock of feed material	Proper microstructure and hence proper properties	Proper microstructure and hence proper properties	Mirror finish (R_a: between 0.5 and 0.7 μm) Compressive residual stresses (between 100 and 150 MPa)	Mirror finish (R_a: between 0.5 and 0.7 μm) Compressive residual stresses (between 100 and 150 MPa)	High wastage	Minimal wastage since lesser stock of feed material is required to begin with. MS_{MANU} = 0.896
	Finish machining to obtain required finish. Depth of cut: 0.7 mm							
	Shot peening to obtain compressive residual stresses							

Table 8.5 (continued)

Case study	Manufacturing process(es)		Properties		Surface integrity		Wastage	
	Conventional	Frugal	Conventional	Frugal	Conventional	Frugal	Conventional	Frugal
Inherently *frugal* approach to *superfinishing* from India (Sooraj and Radhakrishnan 2014)	Use of abrasive and erosion based techniques like *Magnetic Abrasive Finishing* (MAF) and *Fluidized Bed assisted Abrasive Jet Machining* (FB-AJM) respectively for generating ultra fine finish. Expensive due to necessary magnetic field source and other fixtures. Internal finishing on tubes of 440C-58HRC hardened bearing steel	Simple and low-cost setup using reciprocating action of *elastic abrasives* packed against work surface to produce ultra fine finish. Abrasive grains of 23 μm size embedded on elastomeric spheres of ϕ: 2–3 mm. Internal finishing on tubes of 440C-58HRC hardened bearing steel	–	–	R_a between 20 and 30 nm	R_a between 20 and 30 nm	Relatively more material removed due to heavier indenting action of grits. Possible damage to surface	No damage to surface due to flexible elastomeric action resulting in lower penetration and lesser material removal

these parts significantly susceptible to corrosion (Zhang and Attar 2016). Although conventional SLM approach would involve post treatments—such as heat and/or deformation to change columnar grains to equiaxed type—to remedy the grain structure, the underlying invalid phase of acicular martensite, α', is difficult to rectify. Moreover, an extra finishing operation might also be warranted to get the requisite *surface integrity*. In contrast, as shown in Fig. 8.4, frugalizing would entail deposition of Ti-6Al-4V alloy together with an inocoulant on a preheated substrate with the former seeding nuclei for equiaxed microstructure possessing $\alpha + \beta$ phases due to a decrease in cooling rate afforded by the latter. Moreover, a reduced thermal gradient from a preheated substrate would further suppress columnar-grain formation. The creation of a proper grain structure would also facilitate generation of necessary *residual stresses* in the part. However, a light machining-cut would be required to give proper finish and, also proper residual stresses if needed, considering the current limitations of AM. Such limitations would be overcome in the foreseeable future with the current progress being made, not least on surface features of depositions from AM technologies (Ngo et al. 2018; DebRoy et al. 2018). Here again, the *frugal* route does away with the resources and thus costs associated with separate post treatments. In particular, costs would be significantly reduced due to completely avoiding outlays associated with correcting both improper phases and columnar microstructure deposited by conventional SLM. Besides costs, the nearly single-pass nature also results in lower energy consumption, which is crucial since AM is considered to be energy intensive.

The third case study is on machining, which is a workhorse of industry. Advances in plastic flow; self-excited-vibratory phenomenon called chatter (Merritt 1965) and machine-tool systems have lead to significant progress in *high speed machining* (HSM) of metallic materials. As listed under the *frugal* column of Table 8.5, aluminum AL7075 alloy can be machined in a single cut at a high speed of 1350 m/min and a depth of cut of 0.7–2 mm to leave a mirror finish (Ra of 0.5–0.7 μm) accompanied by compressive residual stresses (100–150 MPa) (Rao and Shin 2001). As depicted in Fig. 8.5, such impressive *surface integrity* would warrant multiple passes, corresponding to a thicker stock of feed material and, followed by a post-treatment of shot-peening under conventional machining involving low to moderate speeds. The use of very high cutting speeds in HSM leads to single-pass machining wherein requisite surface integrity, tolerances and properties are obtained in a single cut. So selection of a suitable depth of cut, corresponding to a thinner stock of feed material, under HSM conditions can improve productivity while leading to superior surface integrity and good properties while saving significant amounts of material and also reducing costs. In this regard, an MS_{MANU} of ~0.9 is mentioned in the last column of Table 8.5. Table 8.6 lists details of calculating this MS_{MANU} through Eq. (8.2). The MS_{MANU} of ~0.9 signifies savings of ~90% of Al 7075-T6, by adopting HSM for FM, wasted through conventional machining involving a rough and a finishing pass. Such a high value of MS_{MANU} will feed into Eq. (8.1) and give a commensurately higher value for F of F^S. In other words, suppressing wastage majorly through HSM enhances the frugality credentials of the concerned product. It is noteworthy that other than lesser amount of feed stock, costs in HSM can be further reduced by

Table 8.6 MS_{MANU}: conventional and *frugal* machining of a rectangular block of Al 7075-T6 by uncoated carbide face-milling cutter (volume wasted = 100 mm × 70 mm × depth of cut)

Manufacturing type	Depth of cut (mm)		Volume wasted (mm^3)	MS_{MANU}
	Rough cut	Finish cut		
Conventional	6	0.7	46900	–
Frugal	–	0.7	4900	0.896

employing *minimum quantity lubrication* (MQL) (Tai et al. 2011) wherein a fine mist of concentrated coolant/lubricant is effectively used to significantly influence *surface integrity* and also carry away debris. Moreover, higher speeds lead to more heat generation in the deformation zone which by softening material lead to reduction of cutting forces and hence lesser energy consumption during HSM. Therefore, *high speed machining* is an appealing *frugal* alternative to the majority of conventional machining operations due to the resulting savings in resources, energy and costs.

The case studies covered above are inherently capital intensive. It is noteworthy to mention activities of small-scale manufacturers in India who are inherently *frugal* due to their limited revenues and lower profitability. In this regard, *The Energy and Resources Institute* (TERI) in India has studied a cluster of small-scale component manufacturers that cater to the larger *original equipment manufacturers* (OEMs) associated with the country's thriving automotive sector. TERI has encouraged this cluster in honing their potential for saving materials, reducing emissions and lowering costs. In particular, this cluster of companies in sheet-metal processing has reduced its material wastage by 15%, on average, by simply opting for better ulilization of the ingoing sheet metal so as to reuse the resulting scrap as secondary raw material (TERI 2016). Therefore, the cluster has been able to achieve a cumulative MS_{MANU} of 15% and a systematic application of the approach outlined in this chapter can plausibly wring further savings for the companies while also improving the F^S of their products. Moreover, the manufacturing-research community in India is also focusing on lowering costs through alternate techniques that are better suited for frugality. An example, listed under the last entry of Table 8.5, being the simple technique of superfinishing developed by researchers at the *Indian Institute of Space Science and Technology* (IIST) (Sooraj and Radhakrishnan 2014). The entry in Table 8.5 considers application of superfinishing to hardened bearing steel (440C-58HRC). This technique uses *elastic abrasives*, i.e., elastomeric balls embedded with abrasive grits, for imparting ultra fine finish with minimal penetration of grits thereby saving on material and also not damaging the surface. In particular, the flexible elastomeric spheres confirm to the work surface by pressing against it and in the process provide large surface area with more grits that are under lesser pressure and hence undergo lesser penetration. In contrast, the conventional techniques mentioned in Table 8.5 use expensive magnetic sources and other fixtures to arrive at such a finish with more material removal and possible surface-damage due to the relatively heavier indenting action of grits. Other than simple and low-cost procedures, Indian companies are pro-

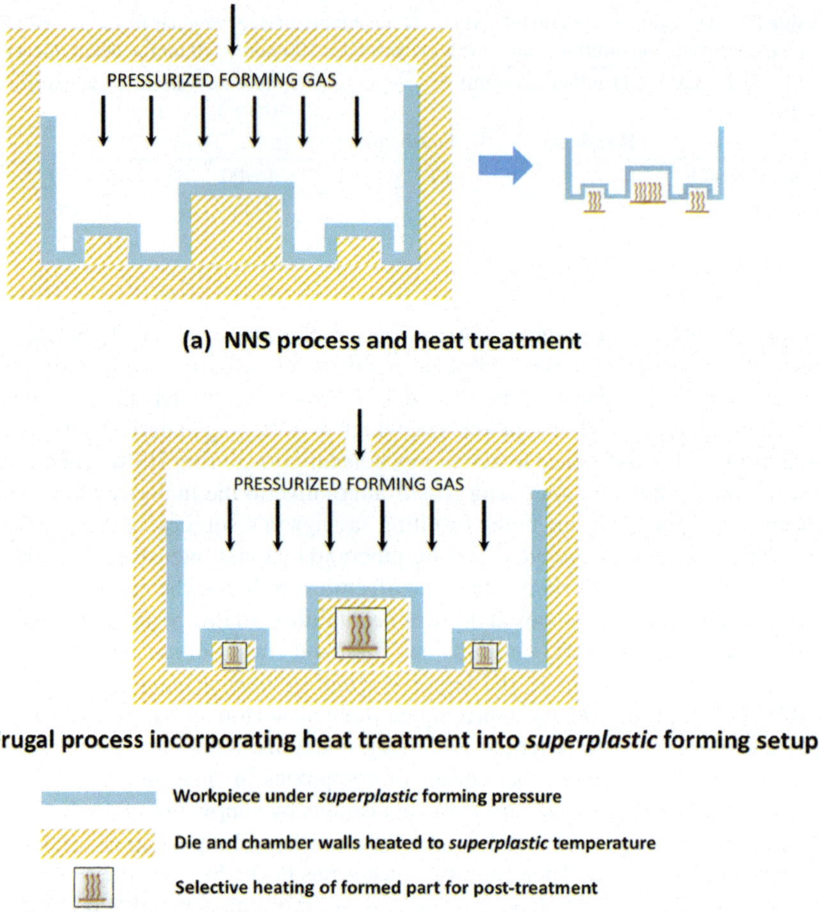

Fig. 8.2 Superplastic forming as a *frugal* process

ducing low-cost machine tools for all kinds of advanced manufacturing operations including those alluded to in the case studies of Table 8.5 (UNIDO 2005). The use of such low-cost machine tools allow considerations of frugality even for advanced manufacturing operations. Overall, such inherently resource-and-cost saving capabilities in India and elsewhere should be tapped into to frugalize manufacturing of any kind, advanced or otherwise.

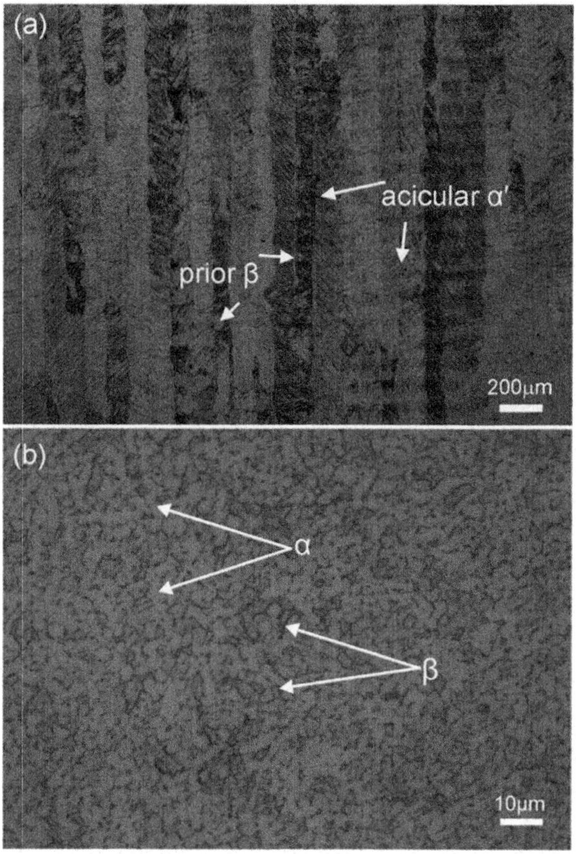

Fig. 8.3 *Selective laser melting* of Ti-6Al-4V alloy results in undesired acicular α' phase along with β in a columnar-grain structure shown in (**a**). In contrast, the cast Ti-6Al-4V alloy in (**b**) possesses the desired $\alpha + \beta$ phases in an equiaxed microstructure. (*Produced with permission from* Zhang and Attar (2016))

8.6 More Insights on Frugal Manufacturing

A *factor of frugality* based approach is a dedicated tool for quantifying frugality in all aspects of design including manufacturing. The savings in resources achieved through MS_{MANU} feeds into improving F^S. Thus, the model for MS_{MANU} though simplistic forms a portion of F^S and as such is crucial to the overall streamlining of a product for *frugality*. Although W_{MAX}, appearing in Eq. (8.2), might refer to hypothetical generation of maximum waste in some instances, its determination helps honing in on the actual *frugal* operation(s) that in turn lead to utmost savings in materials. In other words, selection of most wasteful real-time process(es) for W_{MAX} is justified when prior numbers are unavailable. Moreover, Eq. (8.2) can be used for conserving materials even for manufacturing comprising wasteful operations such as *metal cutting* by aiming for a maximal value of MS_{MANU}. Alternatively, Eq. (8.2) can be used to devise recycling and reuse post treatments necessary for putting such inevitable waste to good use.

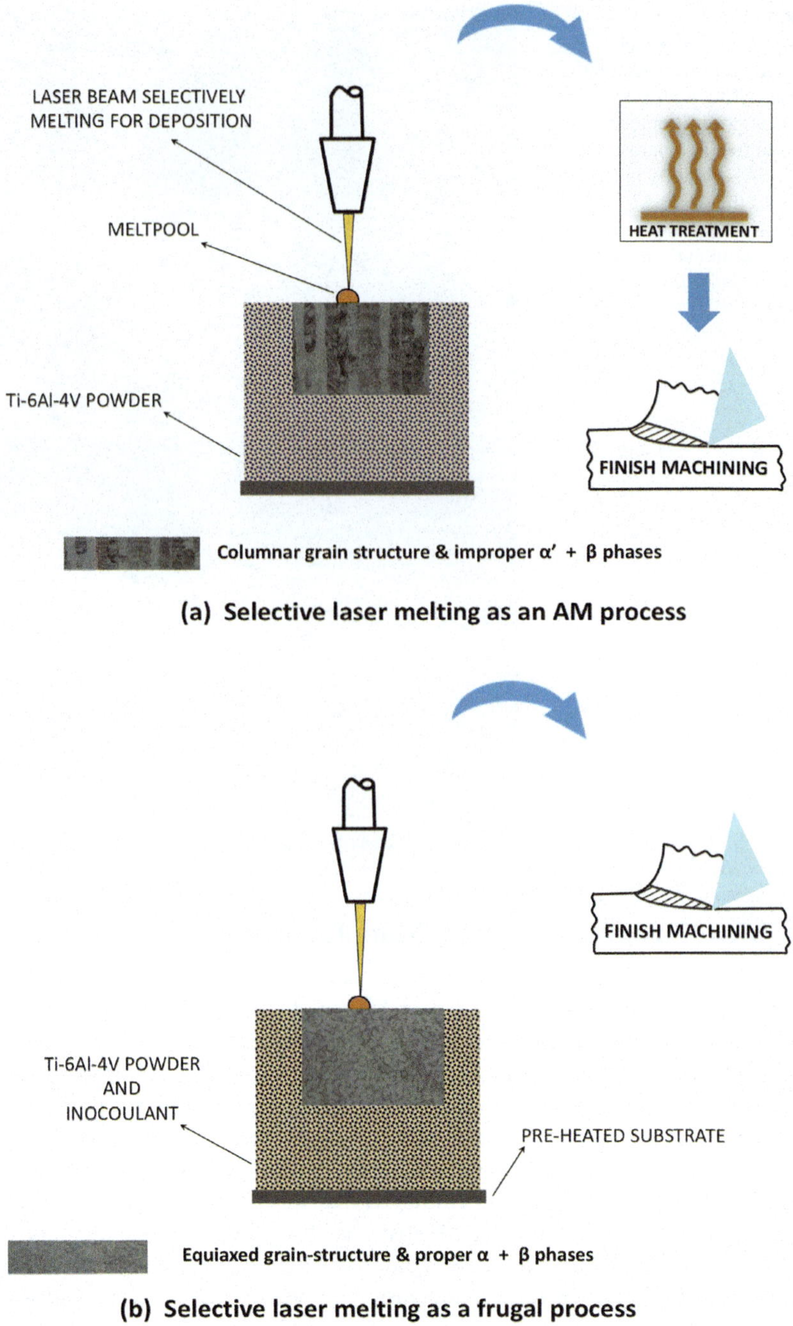

Fig. 8.4 Additive manufacturing as a *frugal* process

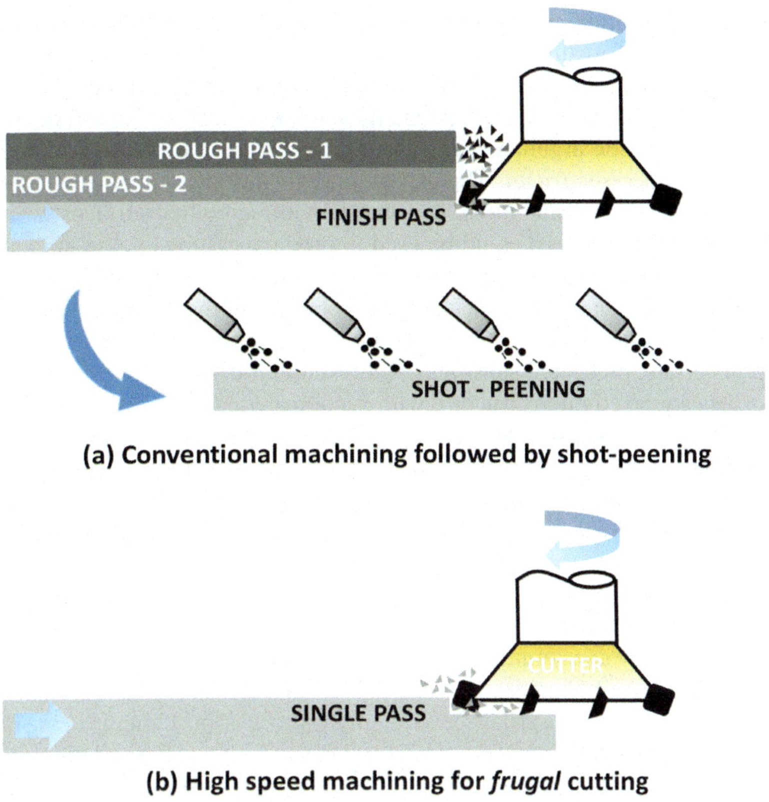

Fig. 8.5 High speed machining as a *frugal* process

The examples of Table 8.5 illustrate the potential of leveraging existing processes for *frugal manufacturing*. Although most of these entries have same wastage (MS_{MANU} values), the *frugal* approach is better in terms of properties and *surface integrity* imparted to the final product with minimal number of operations. These case studies have shown potential for associated cost reductions and energy savings that are significant to fighting *climate change*. Hence, any set of arbitrary manufacturing processes could be utilized to do the same along the lines of Table 8.5. Therefore, manufacturers should ideally adapt all of the existing processes including NNS techniques to being *frugal*. This initial wave of frugalization would also set the tone for discovering newer technologies that are designed for FM at inception. In fact, advances in manufacturing including robotics, automation, AI, big data, *digital twins*, *internet of things* (IoT) and the entire gamut of subtractive and additive techniques would greatly support this endeavor to go *frugal*. Such advances would greatly facilitate achieving highest MS_{MANU} values while aiming for single pass and low-cost energy-saving processes that produce quality products.

Although most of the examples in Table 8.5 have not completely eliminated secondary operations, frugality has been instrumental in reducing their numbers. Therefore, entries of Table 8.5 attest to FM's ability to create single pass processes when non-*frugal* NNS operations have failed in the past (Kudo 1990). Therefore, Table 8.5 lends credence to the arguments accompanying Table 8.2 for making stand alone or individual processes *frugal*. This is possible, for instance, in casting operations where suitable process-conditions and molds with smoother surfaces along with incoulants in the metal feed could be used to produce castings possessing a proper equiaxed microstructure with close tolerances and also finish. Such an equiaxed structure will also lend proper mechanical properties to the casting. *Additive manufacturing* techniques could be utilized with proper process parameters; incoulants in feed materials; preheated substrates and resolution-of-deposition to lay deposits that possess equiaxed microstructure with accompanying superior properties; good *surface integrity*; and hew to dimensional tolerances. The exclusion of *powder metallurgy* techniques from Table 8.2 is due to increasing need for post processing such as sintering treatments to suppress porosity. Another drawback is that sintered PM parts are typically finish-machined to achieve requisite *surface integrity*. Apart from the stand alone mode, the operations of Table 8.2 could also be used to minimize numbers of effective processes and hence make *frugal* any arbitrary set of manufacturing processes. Also, the ideal single-pass nature of FM will generally lead to reduction of energy consumption and, associated emissions.

A special focus of frugalization on *additive manufacturing* is essential. This is due to AM's potential to produce, as a single process in a single pass, intricate shapes with equiaxed microstructure possessing superior mechanical properties that are nearly impossible to fabricate with other traditional or subtractive processes. In this regard, the ability to successfully print equiaxed microstructures for titanium and nickel based alloys (Zhang et al. 2019b; Ning et al. 2018) in a single pass should be extended to other metallic materials. Complex shapes are required for fulfilling modern engineering needs and their printing in one piece is sustainable because of the absence of assembly operations and their associated emissions. The foray of AM into *topology optimization*, wherein material is deposited only in functional regions of a product, also makes 3D printing appealing from the perspectives of both *frugality* and sustainability (Plocher and Panesar 2019). Also, the ability to print large objects like boats in relatively lesser time makes *additive manufacturing* a good alternative to waste-prone traditional processes. Beyond depositing from scratch, techniques such as *laser engineered net shaping* (LENS^{TM}) can be used to repair and hence remanufacture impaired parts (Chandra and Rao 2017) thereby enhancing AM's salvaging and hence *frugal* abilities. Furthermore, AM is leading to novel ways of enhancing product-performance such as printing of a bone-inspired architecture for substantially improving fatigue life of foam-like products (Torres et al. 2019).

Some generalized suggestions for frugalizing the popular operations described in Sect. 3 are listed in Table 8.7. These suggestions are one of several possible solutions for each entry. Also, a given application has to be studied in detail for their frugalization, especially in terms of properties desired in the part/product and, these observations could temper and modify these suggestions. And, the entries of Table 8.7

Table 8.7 Generalized suggestions for frugalizing popular manufacturing processes

Process	Machine-tool system	Work material	Operation	Process conditions
Metal cutting	Low cost	–	Single-pass combining rough and finish cutting	Higher speeds with lowest possible depth of cut
Bulk & Sheet metal	Low cost	–	Single-pass involving requisite treatments. Finish machining only for NNSs	Finishing machining with lowest possible depth of cut
Additive manufacturing	Low cost	Addition of inoculant together with a preheated substrate for achieving proper microstructure	Single build operation followed by finish machining only for NNS-depositions	Finishing machining with lowest possible depth of cut
Casting	Low cost	Addition of inoculant or other techniques for catalyzing proper microstructure	Single pour for molding followed by finish machining only for NNSs	Finishing machining with lowest possible depth of cut
Non traditional methods	Low cost	–	Single-pass followed by finish machining only for NNSs	Finishing machining with lowest possible depth of cut

are amenable to changes corresponding to progress in research on these operations. A common theme is usage of low-cost machine tools and related equipment not compromised in generating products with requisite tolerances, *surface integrity* and other properties. A single operation has also been suggested for all processes with "treatments" referring to retrofitting machines to perform any post treatments in tandem. As for *metal cutting*, higher speeds have been suggested along with single tool-passes for imparting better properties to fabricated surfaces.

The tenets of FM should be utilized to build the next generation of machine tools. However, development of *frugal machine-tool systems* (FMTSs) would have to overcome challenges rooted in the physics of manufacturing operations. An instance would be the production of NFSs through single passes of traditional forming processes possessing attendant *frugal* features encountering, inter alia, higher loads arising from the precision required of the operations (Kudo 1990). Also, high quality feed stock for *bulk* and *sheet* operations (Kudo 1990) is a must to minimize defect rates and hence facilitate successful generation of NFSs. Hence, relevant mechanics underlying existing manufacturing operations together with current advances should be utilized to build FMTSs from scratch or retrofit existing machine tools. In this regard, tools of Industry 4.0, such as big data, digital twins, AI etc., can be leveraged in creating FMTSs. Consequently, large outlays incurred in developing FMTSs can be recovered during usage of these machine tools due to cost savings from streamlined manufacturing operations.

Although assembly operations are ideally avoided in FM, their need in real time where necessary should be kept to a minimum. For some products, especially complicated type such as commercial aircrafts, assembly of fabricated parts is crucial for completion of the final product. In this regard, parts designed for frugality (Rao 2017b, 2019) and made from FM processes can be finally assembled into a given product. Therefore, until manufacturing evolves to being low-cost single pass operations delivering proper properties and also *surface integrity*, assembly should be designed for frugality.

Frugal manufacturing should be universally applied to metals, non-metals and composites. This is because production technologies possessing climate-friendly credentials are significant to producers of any stripe. Consequently, the *frugal* framework can be adopted for non-metallic materials by considering relevant behavior and also mechanics of deformation process(es) involved in their production. Accordingly, properties, *surface integrity*, shape accuracy, single pass and low cost have to be defined and selected based on the microstructure and behavior of such materials. *3D printing* is well suited to FM of non-metallic materials and its success has been documented in recent years in depositing polymeric and also polymer-like materials for a wide range of applications including biomedical, aerospace, automotive, toy-manufacturing and others (Ngo et al. 2018). In fact, AM techniques are currently well established for polymeric and also polymer-like materials—with an intricate geometry—that do satisfy the tenets of frugality by producing parts in a single pass with good quality accompanied by minimal wastage at low cost (Song et al. 2017; Ngo et al. 2018). Notwithstanding some issues of anisotropy, *3D printed* plastics show improved mechanical properties in certain directions vis-à-vis the traditionally injection-molded parts (Song et al. 2017). Even ceramics and carbon-fiber-reinforced-polymer-composites have been *3D printed* recently into intricate geometries (DeNardo 2016; Halloran 2016).

Frugal manufacturing combined with the pillars of *smart factories* would buttress the foundations of sustainability. In addition to *smart factories'* tools easing adoption of frugality, the ensuing *frugal* activities would be environmentally benign and also produce quality affordable products with improved life. Therefore, futuristic *smart factories* saving on resources, energy and costs through F^S would make manufacturing truly *smart*. Moreover, *smart factories'* tools and techniques would aid in designing *frugal* operations for even small manufacturers whose budgets are limited.

Apart from being rooted in sustainability, *smart factories* incorporating frugality can also have unforeseen applications as in pandemics such as the current one involving COVID-19. In particular, FM's need for lesser numbers of workers due to its typical single-process-with-a-single-pass nature would along with amenities of *smart factories* go a long way in fighting pandemics. This is because first, the absence of secondary operations would by itself lessen opportunities for maintenance, breakdowns and other related services. And second, a smaller workforce together with cleaner facilities due to frugality and other modern technologies would avoid congestion of people and hence danger of contagion.

The starting framework being first generation covers various aspects of manufacturing and as such results in generalizations, among other limitations, which need to be addressed in subsequent efforts. The models developed are first generation that while adhering to the tenets of FM are simplistic. These models could be revised to better capture the intricacies of manufacturing. Moreover, qualitative observations need to make way for quantification as FM is adopted in specific applications. Such quantification will also involve characterization of ensuing savings in energy and costs. Also, FM aims for savings of any kind during manufacturing and some of the indirect material-savings coming from, for instance, improvements in *surface integrity* might not be high in some cases. However, cumulative savings from a bulk of such applications would be appreciable. Similarly, smaller values of MS_{MANU} may seem insignificant but their contribution over many products would give higher savings in materials and also costs.

Utmost reduction of resource consumption is crucial for all round *sustainable development* and manufacturing is an important activity for effecting this change. This chapter has presented *frugal manufacturing* for generating zero waste while aiming for better part-quality in terms of *surface integrity* and also mechanical and non-mechanical properties. *Frugal manufacturing* also entails use of single-pass low-cost manufacturing processes for achieving these goals. These challenging demands can be met by quantifying manufacturing-frugality through a recently developed design approach based on the *factor of frugality* (F^S), a modern version of the *classical factor of safety* (S). Real time frugalization examples have been presented on machining of an aluminum alloy and superplatic forming and selective laser deposition of nickel- and titanium-based superalloys respectively. Also, the inherently Indian ability of frugality has been exemplified by superfinishing of a hardened steel. Such low-cost processes and machine-tool systems can buttress the low-cost feature of *frugal manufacturing*. The potential savings in resources, energy, emissions and hence costs demonstrated by these examples, while using a minimum number of processes for producing quality parts, should be harnessed by blanket application of *frugal manufacturing*. Other than designing new-age machine-tool systems for existing and newer processes, *smart factories* should also get a fillip by harnessing frugality. By incorporating *frugal manufacturing* as another pillar, *smart factories'* efficacy in making manufacturing green would be enhanced. Moreover, a smaller workforce due to the single-process-with-single-pass nature of *frugal manufaturing* would even aid in fighting pandemics such as the current one involving COVID-19.

Questions

The questions to this chapter assume that the reader is well versed with manufacturing technology and processes. If not, the motivated reader is encouraged to complete the undergraduate course on the same. And, most of the problems in this section are open ended and the reader should not hesitate to delve into the details for answering them. These are also sample problems, showing hopefully a reasonable variety of ways to make manufacturing *frugal* and, engineers, scientists and other readers are encouraged to look at various fabrication, assembly and machine-tool design problems from the angle of frugality. These will lead to improvements which with time can benefit manufacturing immensely.

1. The *grinding* process, shown in Fig. 8.6, is an abrasive manufacturing operation that removes material by abrasive action of grains bonded together in a rotating wheel. These abrasive grains do not have cutting edges with a fixed geometry as in machining but are randomly oriented to remove material under rotary motion against the horizontal feed of the workpiece. The "b" portion of this figure shows the section of a rectangular slab of a metallic alloy whose thickness "t" has to be removed by *metal cutting*. However, the resulting surface needs to have a super finish with compressive residual stresses. So,
   a. Can *grinding* be employed as a *frugal* process?
   b. Explain the answer to the above question using the tenets of *frugal manufacturing*.

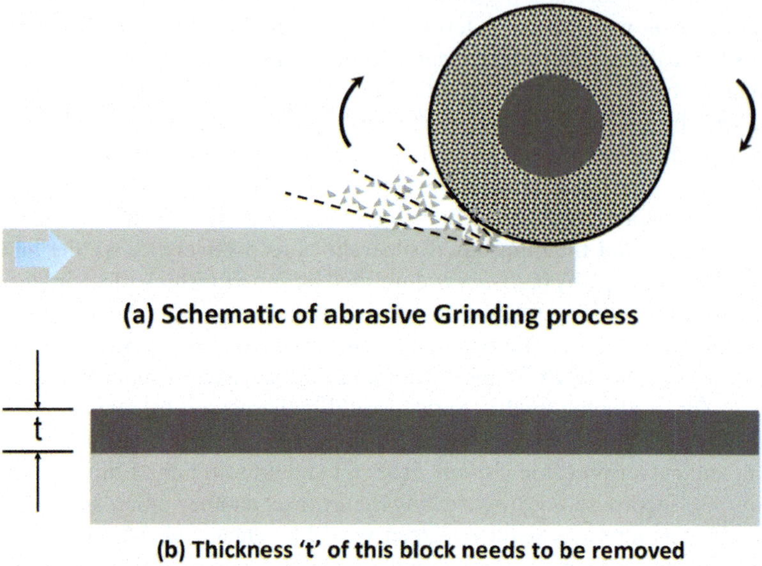

Fig. 8.6 The *grinding* process and the workpiece whose thickness has to be reduced

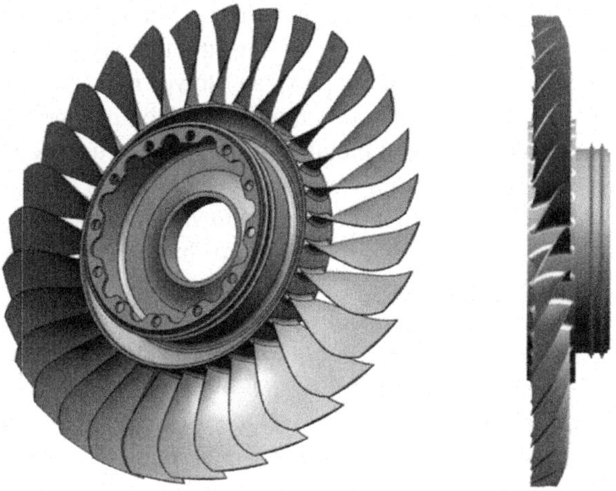

Fig. 8.7 A typical compressor rotor of aircraft jet-engine manufacture by *blisk* technology. Schematic taken from Fricke et al. (2021)

Hint: Answers to this and the next question are more a function of manufacturing process(es) employed and dependence on a particular metallic material can be neglected in a first approximation to answers of these questions.

2. Manufacturing processes are also energy intensive. So does your answer to question (1) remain the same from a perspective of energy consumed during manufacturing of the metallic block? If not explain alternative(s) necessary for this operation.
3. The aviation industry is a pinnacle of *frugal engineering* wherein overall design of an aircraft uses a low *factor of safety* closer to unity. This economizing of design is needed for minimizing fuel consumption (also important from *climate change* point of view). However, the design and manufacturing of individual parts can still be improved to economize on other fronts. The *factor of frugality* can be very useful to the aerospace industry in attempting to wring out further savings in resources, cost and possibly leading to improvements in fuel efficiency. A critical jet-engine case study selected for this question concerns manufacturing of rotor disks with blades, shown in Fig. 8.7, using *blisk* (blade integrated disks) technology. In the *blisk* process, used for rotors with larger blades, the disks are machined and subsequently friction welded to blades forged individually. Answer the following:
   a. Analyze the blisk process and compute the *factor of frugality* accounting for every stage of product development.
   b. Explore improving the manufacturing of these rotor-with-blades by the *blisk* process using the knowledge garnered while answering the above question.
   c. Are there alternative(s) to *blisk* based on your *frugal* analysis and if so what are they?
4. List manufacturing process(es) individually or in combination, other than those presented in this chapter, that could be used for *frugal manufacturing*. These process(es) could be used directly (i.e., as it is) or modified for this purpose.

5. There is a strong relationship between production rate (or number of parts produced) and cost of a part. Discuss this relationship from a *frugal manufacturing* perspective when quality and lower resources also have to be accounted for in realizing the *frugal* version of a given part/product.
6. In the casting of metallic materials, discuss the influence of material; part shape and size; accuracy; equipment; and labor on the cost of a cast product from a viewpoint of *frugal manufacturing*.
7. Explain the importance of hot hardness to metal cutting operations in *frugal manufacturing*.
8. *Parasitic mechanisms* in manufacturing dissipate energy that is not utilized in the fabrication of the component/product. These mechanisms hamper waste-and-cost reduction and also quality, all of which are of importance in *frugal manufacturing*. Study *parasitic mechanisms* of manufacturing operations in general and list operations with the most number of such mechanisms.
9. Explain the *parasitic mechanism* of chatter in metal cutting operations and explain the importance of its reduction in *frugal manufacturing*.
10. A cylindrical workpiece of Aluminum alloy is mounted on a machine tool and which is 1.5 ft. long with a diameter of 1.5 in. The slender nature of the workpiece together with the dynamics of the machine tool make the overall system prone to the self-exciting vibration phenomenon of chatter. This workpiece needs to be longitudinally turned with a tool having a positive rake angle of 5° at some RPM and a feed of 0.05 mm/rev. Engineers A and B at *FrugalTech* company have this data but B is also cognizant about *frugal manufacturing* and accordingly has access to a *stability lobe diagram* (SLD) for this problem. This SLD is shown in Fig. 8.8. Engineer A is however just relying on his "experience" and will experiment with cutting in arriving at suitable cutting conditions. An optimal depth of cut has to be used at a suitable RPM for arriving at the maximum *metal removal rate* (MRR) without chatter. Explain which engineer will generate lesser waste for the company in a planned manner and how through suitable calculations for determining MRR and machining time for the data given in this question. You have all the necessary data to make a proper decision!
11. Hot, warm and cold working correspond to different ranges of *homologous temperature* (T_T). How is hot, warm and cold working used to advantage in bulk deformation processes? Explain which or none of these working conditions would be beneficial in the *frugal manufacturing* of bulk metallic products.
12. The defects in products during their manufacturing can arise in the material used and the process employed for shaping the material. Assuming that material-base defects are absent, list the defects for the following processes:
    a. Casting
    b. Forging
    c. Rolling
    d. Extrusion
    e. Sheet-metal operations
    f. Metal cutting

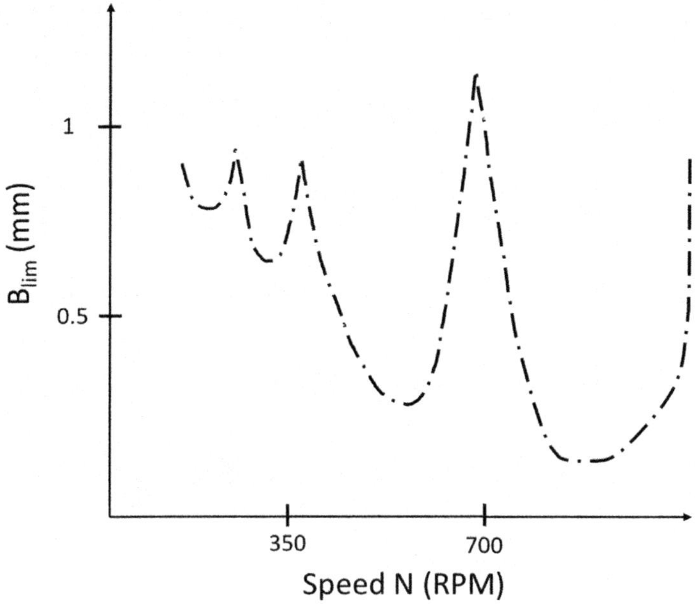

Fig. 8.8 Stability lobe diagram for longitudinal turning of an Aluminum alloy

13. Are the process-based defects of the previous question important from a *frugal manufacturing* perspective? Explain your answer.
14. List some material-based defects independent of any process and explain their importance to *frugal manufacturing*
15. Extrusion of metallic materials can be carried out by the direct or indirect process, see Fig. 8.9. Which of these is better for *frugal manufacturing* of metallic materials and why?
16. An extrusion process has to be designed for a workpiece made of stainless-steel. This workpiece has to be reduced in diameter from 110 to 55 mm. The extrusion process can be carried out hot or cold and this has a huge influence on the force-magnitude and hence affects selection of a press with the proper tonnage. Based on this information select a suitable press with proper tonnage and explain why it qualifies for *frugal manufacturing*.
17. *Forming limit diagrams* (FLDs) are indispensable to manufacturing of sheet-metal parts. Explain how FLDs are also indispensable to the *frugal manufacturing* of sheet-metal parts.
18. Any manufacturing operation comprises active and idle portions wherein the former accounts for actual fabrication while latter accounts for the preparatory actions necessary to get the operation ready and also tool-work-positioning during the operation. In other words, idling accounts for any non-fabrication portion necessary to carry out the operation. An engineer can reduce idle time (ideally to zero) by controlling tool-workpiece geometry and parameters appear-

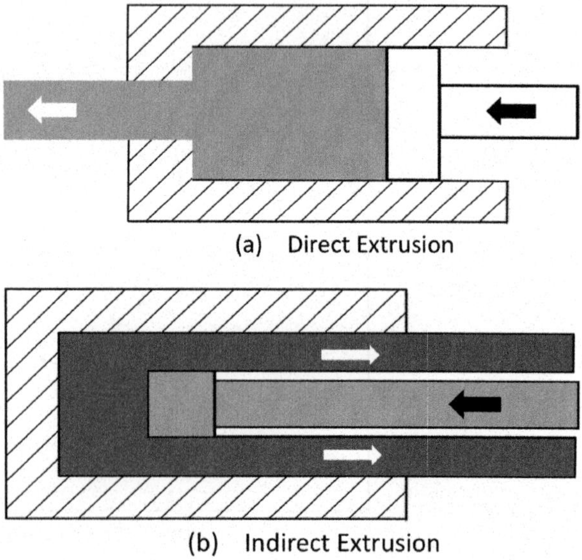

Fig. 8.9 The direct and indirect routes to extrusion of metallic materials

ing in formulations for the *metal removal rate* (MRR), fabrication-time and other process-specific models. Such reduction of idling translates to lower costs and hence make manufacturing *frugal*. Describe your models for maximum reduction of idling in the following processes (pick any type in each) that will be significant to designing appropriate *frugal machine-tool systems*:

a. Casting
b. Forging
c. Rolling
d. Extrusion
e. Sheet-metal operations
f. Metal cutting

19. Low-cost aspect of *frugal manufacturing* is also associated with energy savings. In other words, energy savings in a process typically translate into lower expenditures for a given price of available energy. Mention *metal cutting* operations that are energy intensive and devise methods for making these operations energy efficient. **Hint: Use phenomena such as size-effect in *metal cutting* energy consumption shown in Fig.** 8.10.
20. Repeat the solution to the previous question for all other manufacturing processes, i.e., bulk, sheet-metal and additive.
21. The aviation industry overall employs lower *factors of safety* in designing aircrafts with lower weights. Therefore, commercial aircrafts, a representative shown in Fig. 8.11, can be considered as *advanced frugal innovations* whose engineering has been streamlined plausibly to a high degree. Is there a possibility of wringing more savings in materials and costs by considering *frugal engineering* principles?

    Hint: This question needs the reader to utilize all pillars of *frugal engineer-*

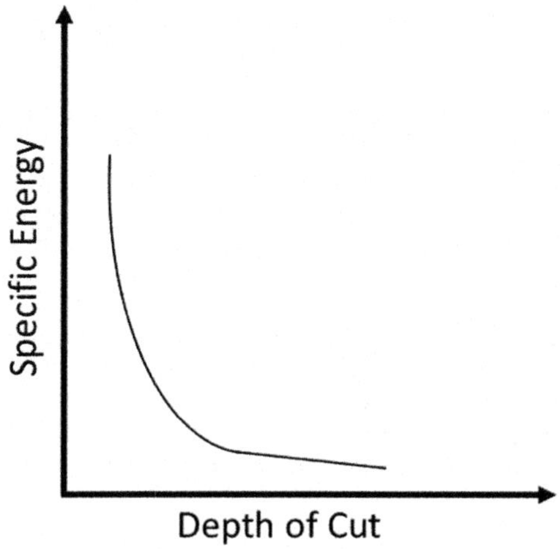

Fig. 8.10 Variation of energy expended per unit volume of material removed in *metal cutting*. This clearly shows the size-effect at very low depths of cut

Fig. 8.11 A representative commercial aircraft used for long distance flights. Image downloaded from https://pixabay.com

***ing* covered upto and including this chapter. In other words, use of the *factor of frugality* would help the reader to systematically comb for more savings in various steps (in correct sequence) of product development.**

References

Altan T, Miller RA (1990) Design for forming and other near net shape manufacturing processes. Ann CIRP 39(2):609–620

Bassini E, Iannucci L, Lombardi M, Biamino S, Ugues D, Vallillo G, Picqué B (2019) Net shape hipping of a ni-superalloy: a study of the influence of an as- leached surface on mechanical properties. J Mater Process Technol 271:476–487

Benedict GF (1987) Nontraditional manufacturing processes. Marcel Dekker, New York, USA

Bewlay BP, Gigliotti MFX, Hardwicke CU, Kaibyshev OA, Utyashev FZ, Salischevb GA (2003) Net-shape manufacturing of aircraft engine disks by roll forming and hot die forging. J Mater Process Technol 135:324–329

Chandra S, Rao BC (2017) A study of process parameters on workpiece anisotropy in the laser engineered net shaping (lenstm) process. J Phys D: Appl Phys 50(22):225303

Cominotti R, Gentili E (2008) Near net shape technology: an innovative opportunity for the automotive industry. Robot Comput Integr Manuf 24:722–727

Cook AJ, Werner PS (1991) Pressure infiltration casting of metal matrix composites. Mater Sci Eng A 144:189–206

DebRoy T, Wei HL, Zuback JS, Mukherjee T, Elmer JW, Milewski JO, Beese AM, Wilson-Heid A, De A, Zhang W (2018) Additive manufacturing of metallic components—process, structure and properties. Progress Mater Sci 92:112–224

DeNardo NM (2016) Additive manufacturing of carbon fiber-reinforced thermoplastic composites. Master's thesis, Purdue University

EPA (2015) Sustainable manufacturing. https://archive.epa.gov/sustainablemanufacturing/web/html/business-case.html

Fricke K, Gierlings S, Ganser P, Venek T, Bergs T (2021) Geometry model and approach for future blisk lca. IOP Conf Ser: Mater Sci Eng 1024:012067

Gilissen R, Erauw JP, Smolders A, Vanswijgenhoven E, Luyten J (2000) Gelcasting, a near net shape technique. Mater Des 21(4):251–257

Halloran JW (2016) Ceramic stereolithography: additive manufacturing for ceramics by photopolymerization. Ann Rev Mater Res 46:19–40

Herderick E (2011) Additive manufacturing of metals: a review. ASM

Hoekstra AY, Wiedmann TO (2014) Humanity's unsustainable environmental footprint. Science 344:1114–1117

Kang CG, Jung YJ, Youn SW (2003) Horizontal reheating of aluminum alloys and semi-solid casting for a near net shape compressor component. J Mater Process Technol 135:158–171

Knorringa P, Pesa I, Leliveld A, van Beers C (2016) Frugal innovation and development: aides or adversaries? Eur J Dev Res 28(2):143–153

Kudo H (1990) Towards net-shape forming. J Mater Process Technol 22:307–342

Lewandowski JJ, Seifi M (2016) Metal additive manufacturing: a review of mechanical properties. Ann Rev Mater Res 46(1):151–186. https://doi.org/10.1146/annurev-matsci-070115-032024

Liefner I, Losacker S, Rao BC (2020) Scale up advanced frugal design principles. Nat Sustain. https://doi.org/10.1038/s41893-020-0585-7

Liu Y, Xu X (2017) Industry 4.0 and cloud manufacturing: a comparative analysis. J Manuf Sci Eng 139:034701

Marini D, Corney JR (2017) A methodology for near net shape process feasibility assessment. Prod Manuf Res 5(1):390–409

Marini D, Cunningham D, Corney JR (2017) Near net shape manufacturing of metal: a review of approaches and their evolutions. Proc Inst Mech Eng Part B: J Eng Manuf 232(4):650–669

Merritt HE (1965) Theory of self-excited machine-tool chatter: contribution to machine-tool chatter research–1. ASME J Eng Indus 87(4):447–454

Miesler T, Wimschneider C, Brem A, Meinel L (2020) Frugal innovation for point-of-care diagnostics controlling outbreaks and epidemics. ACS Biomater Sci Eng 6(5):2709–2725

Moore JB, Tequesta J, Athey RL (1970) Fabrication method for the high temperature alloys. http://www.google.it/patents/US4741207. US Patent 3,519,503
Morita A, Hattori S, Tani K, Takemura A, Ashida Y (1991) Near net shape forging of titanium alloy turbine blade. ISIJ Int 31(8):827–833
NAS (2019) Environmental engineering for the 21st century: Addressing grand challenges. Technical Report, National Academies of Sciences, Engineering, and Medicine. DC, USA, Washington
Nature (2014) Editorials: renewed energy. Nature 510:7–8
Ngo TD, Kashani A, Imbalzano G, Nguyen KTQ, Hui D (2018) Additive manufacturing (3d printing): a review of materials, methods, applications and challenges. Compos Part B Eng 143:172–196
Ning FD, Hu YB, Liu ZC, Wang XL, Li YZ, Cong WL (2018) Ultrasonic vibration-assisted laser engineered net shaping of inconel 718 parts: microstructural and mechanical characterization. J Manuf Sci Eng ASME 140(6):061012
NSTC (2013) National network for manufacturing innovation: a preliminary design. Technical report, NSTC: Advanced Manufacturing National Program Office
Onodera S, Sawai K (1992) Current cold-forging techniques for the manufacture of complex precision near-net-shapes. J Mater Process Technol 35:385–396
Park JW, Lee JG, Jun CS (2009) Near net-shape five-axis face milling of marine propellers. Int J Precis Eng Manuf 10(4):5–12
Plocher J, Panesar A (2019) Review on design and structural optimisation in additive manufacturing: towards next-generation lightweight structures. Mater Des 183(5):108164
Prabhu JC (2017) Frugal innovation: doing more with less for more. Philos Trans Roy Soc A: Math Phys Eng Sci A375:20160372. https://doi.org/10.1098/rsta.2016.0372
Rao BC (2019) The science underlying frugal innovations should not be frugal. Roy Soc Open Sci 6(5). https://doi.org/10.1098/rsos.180421
Rao BC, Shin YC (2001) Analysis on high-speed face-milling of 7075-t6 aluminum using carbide and diamond cutters. Int J Mach Tools Manuf 41:1763–1781
Rao BC (2013) How disruptive is frugal? Technol Soc 35:65–73
Rao BC (2017) Revisiting classical design in engineering from a perspective of frugality. Heliyon 3(5):e00299
Rao BC (2017) Advances in science and technology through frugality. IEEE Eng Manage Rev 45(1):32–38
Rao Balkrishna C (2020) On complex systems of adaptive frugal products. Roy Soc Open Sci 7:192057. https://doi.org/10.1098/rsos.192057
Saldana C, Swaminathan S, Brown T, Moscoso W, Mann J, Compton W, Chandrasekar S (2010) Unusual applications of machining: controlled nanostructuring of materials and surfaces. ASME: J Manuf Sci Eng 132(3):030908. https://doi.org/10.1115/1.4001665
Sherby OD, Wadsworth J (1985) Superplasticity and superplastic forming processes. Mater Sci Technol 1:925–936
Song Y, Li Y, Song W, Yee K, Lee K-Y, Tagarielli VL (2017) Measurements of the mechanical response of unidirectional 3d-printed pla. Mater Des 123(5):154–164
Sooraj VS, Radhakrishnan V (2014) Fine finishing of internal surfaces using elastic abrasives. Int J Mach Tools Manuf 78:30–40
Tai BL, Dasch JM, Shih AJ (2011) Evaluation and comparison of lubricant properties in minimum quantity lubrication machining. Mach Sci Technol 15:376–391
Tateno M (1985) Development of large size high quality steels and their future prospects as "near net shape" material. Trans Iron Steel Inst Japan 25:97–108
TERI (2016) Towards a resource efficient auto component manufacturing in India: a case study. Workshop on towards a resource efficient auto component manufacturing in India
Tiwari R, Herstatt C (2012) Open global innovation networks as enablers of frugal innovation: propositions based on evidence from India. Hamburg, Germany. Institute for technology and innovation management working paper, Hamburg, Germany

Torres AM, Trikanad AA, Aubin CA, Lambers FM, Luna M, Rimnac CM, Zavattieri P, Hernandez CJ (2019) Bone-inspired microarchitectures achieve enhanced fatigue life. Proceedings of the National Academy of Sciences (PNAS) of the United States of America

Ugural AC (2015) Mechanical design of machine components. CRC Press, New York, USA

UNIDO (2005) National programme for development of the machine tool industry in India: a success story. Technical report, United Nations Industrial Development Organization (UNIDO), Vienna

Wilde G (2014) Chapter 26: physical metallurgy of nanocrystalline metals. Elsevier, the Netherlands

Withers PJ (2007) Residual stress and its role in failure. Rep Progress Phys 70:2211–2264

Witulski T, HeuBen JMM, Winkelmann A, Hirt G, Kopp R (1994) Near net shape forming of particulate reinforced al-alloys by isothermal forming compared to semi solid forming. J Mater Process Technol 45:415–420

Yamagata H (2005) The science and technology of materials in automotive engines. CRC Press, New York, USA

Yamamoto Y, Kiggans JO, Clark MB, Nunn SD, Sabau AS, Peter WH (2010) Consolidation process in near net shape manufacturing of Armstrong cp-ti/ti-6al-4v powders. Key Eng Mater 436:103–111

Zeschky M, Widenmayer B, Gassmann O (2011) Frugal innovation in emerging markets: the case of Mettler Toledo. Res-Technol Manage 54(4):38–45

Zhang Z et al (2019) Residual stress distribution in ni-based superalloy turbine discs during fabrication evaluated by neutron/x-ray diffraction measurement and thermomechanical simulation. Mater Des 166:107603

Zhang LC, Attar H (2016) Selective laser melting of titanium alloys and titanium matrix composites for biomedical applications: a review. Adv Eng Mater 18(4):463–475

Zhang D, Qiu D, Gibson MA, Zheng Y, Fraser HL, StJohn DH, Easton MA (2019) Additive manufacturing of ultrafine-grained high-strength titanium alloys. Nature 576:91–95

Chapter 9
Quality Control of Frugal Products

9.1 Introduction

The advent of frugal innovation and its subsequent widespread applications are a testimony to humankind's ability to leverage scarcity based constraints for effectively innovating on a global scale. Frugal innovations are increasingly being applied to improve living standards of society at large (Zeschky et al. 2011; Tiwari and Herstatt 2012; Rao 2013; Knorringa et al. 2016; Prabhu 2017; Agarwal et al. 2017; Tiwari and Bergmann 2018). In this regard, witness the increasing numbers of *grassroots* and *advanced* products such as MittiCool and GravityLight, respectively, being utilized in rich and poor segments alike (Rao 2017a). Their appealing list of features including no-frills structure, robust functionality and low cost are also endearing these innovations to practitioners of sustainability. In other words, their "sustainable" appeal stems from catering to society at large with affordable products that are frugal in resource consumption while functioning robustly under various scenarios of the operating-life of these products. Therefore, widespread deployment of frugal products can significantly buttress efforts at sustainably improving living standards of people in all strata by offering quality products that are not expensive.

However, the features making frugal innovations suitable for sustainable development also adversely affect their creation and subsequently demand stringent quality control. In particular, the no-frills structure typical of these innovations neccessiate undertaking of rigorous methodologies for their design (Rao 2019). Methodology for the systematic rigorous design of frugal products have been reported (Rao 2017b) with need for studies on controlling and also verifying the quality of the resulting products' performance. Other than rigorous design, it is also imperative to do rigorous quality checks on the performance of frugal products.

Overall, frugal innovations or frugal products are catching up technologically to serve the needs of humankind. Their features including minimal material consumption, affordability and good functionality make them green from the point of view of sustainable development. However, their no-frills structure typically also necessiates application of rigorous procedures for their design. This chapter outlines the

© The Author(s), under exclusive license to Springer Nature Singapore Pte Ltd. 2024

B. C. Rao, *Frugal Engineering*, Design Science and Innovation,
https://doi.org/10.1007/978-981-99-9700-8_9

importance of quality of frugal products generally in terms of robust functionality under various scenarios of real-time loading. Accordingly, importance of adherence to rigorous design procedures in the factor of frugality approach and also good performance will be emphasized. In particular, various aspects of achieving rigor in design with scarce resources together with some traditional concepts of quality control for extending them to the realm of frugal products will be covered. The quality checks would aid in developing long term competitive solutions for frugal products while the control would facilitate adherence to stringent product specifications. Accordingly, this chapter stresses the breaking up of quality check into two parts for confirming both the rigor of design procedures and also performance of resulting frugal products. It should be noted that this chapter is by no means a comprehensive guide to quality issues for frugal products. It will allow the reader to gain insight into the importance of quality for frugal products and ramifications on how quality would affect practices of purveyors of these products. Besides, most of the content is an extension of the *classical* work by Juran et al. (1974) to account for features unique to frugal products.

9.2 Design and Quality

The frugal consumption of resources is generally a major constraint while realizing products that deliver performance at par with traditional or conservative products that are padded with excess material. The resulting typical lower factors of safety hover around failure conditions making it imperative to undertake rigorous design procedures and subsequent stringent quality checks to confirm the necessary robust functionality of frugal products. The design procedures necessary for frugality have been published by the author (Rao 2017b) with more research updating this effort in the foreseeable future. The sub-sections to follow describe briefly the developed frugal design approach and also quality of developed products, the theme of this chapter.

9.2.1 *Frugal Design Approach*

The *frugal design approach* (FDA) based on the *factor of frugality* (F^S) has been reported recently in Rao (2017b, 2019). Accordingly, this sub-section is based on the design-procedure documented in detail in Rao (2017b, 2019). Here F denotes the *factor of frugality*, which subsumes the *classical factor of safety* S. In other words, F accounts for the scientific principles of traditional design through S, which will have a low value of $\sim$1.5 for *frugal* products. The low value of *safety factor* offsets padding of excess materials thereby making the design light-weight but also typically “flimsy”. Therefore, rigorous design procedures have to be methodically followed for achieving a low S design that saves on materials but also possesses robust functionality. The rigor in design is also essential to get-it-right in the conceptual or design stage since any problems later in the product-development process could increase costs. Such a rigorously designed low *S* design consuming lesser materials is made to loose more weight through other extraneous schemes under *Material Saved* (MS). The *MS* schemes pertaining to various aspects of design including simplicity-in-design, materials, biomimetics and others are utilized to skimp on material in excess to that saved by a low *S* design. Therefore, the *factor of frugality* is given by,

$$F = S + \sum_{i=1}^{n} MS_i \tag{9.1}$$

where, the index *i* indexes *MS* through its *n* components corresponding to *n* extraneous schemes. Equation (9.1) is a first generation model details of which have already been published in Rao (2017b, 2019). The interested reader can refer to these efforts and also a chapter to appear in an forthcoming book on *capturing frugal innovations* (?) for delving deeper into FDA.

9.2.2 Quality of Frugal Products

The frugal products designed and also fabricated according to the tenets of FDA have to go through a quality check. This is because a quality check confirms the theoretical credentials of a given product and whose absence for frugal products would not confirm performance, thereby allowing diffusion of improper products. And this is more so for frugal products, whose origins depend on scarcer resources and, thus warrant thorough scrutiny of their quality. Although quality is looked upon as "fitness-for-use" by consumer (Juran et al. 1974), the "conformance-to-specifications" for makers (Juran et al. 1974) of frugal products entails serious efforts due to the frugal nature of these products. In other words, the onus of creating quality is on makers of frugal innovations since customer is not aware of the frugal nature of a product but expects excellent functionality. And this responsibility can be well managed by proper policies and mission statements by makers to coordinate activities by all stakeholders including design, manufacturing, sales and finance for the success of their frugal products.

Another reason for stressing on proper quality control of frugal products stems from the costs involved. First customers have to be wooed with lower prices. And second, purveyors typically will look forward to reining in their costs in making these products. Proper quality control of frugal products can in this regard facilitate reconciling these conflicting demands. Also, quality control is important for frugal products, in the first place, because of the need to ideally avoid costs associated with repairing and also mending goodwill after releasing bad quality products into the market.

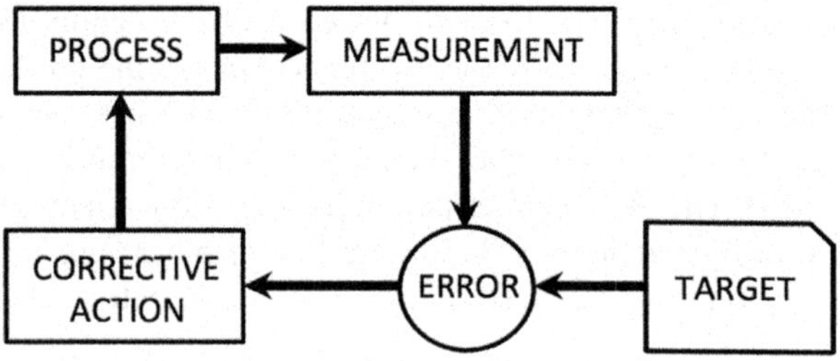

Fig. 9.1 Feedback loop expected of every process contributing to making of a frugal product. Adapted from (Juran et al. 1974)

The quality of a frugal product should be checked in two stages. First, it is imperative to confirm strict adherence to the tenets of FDA. This is because good performance of frugal products will not result without the rigor that is essential to their design. Any deviations from the established rules of FDA would possibly compromise the integrity of product. The confirmation of adherence to rigorous FDA procedures will pave the way for the second stage involving vetting the actual product performance. The performance refers to the service or application rendered to the user whose positive experience would be contingent on the frugal product having successfully cleared the second stage.

9.3 Control of Quality

In extension of *classical* principles, any activity planned for its quality control has to first be put under *statistical control*. This is because statistical procedures are a must for any quality related programs because statistical measures aptly quantify nominal values and the accompanying variance of a characteristic of that activity. Accordingly, this section assuming prior implementation of statistical control, presents only some aspects of quality control that would be significant for frugal products.

The actual quantification of quality of a frugal product is possible through the traditional principles of quality control (Juran et al. 1974). Figure 9.1 shows a schematic of quality control as practiced generally (Juran et al. 1974). Although Fig. 9.1 shows a general schematic, every process inherent to making frugal products, including engineering and non-engineering, should be stringently controlled to achieve the high precision "targets" required for successful frugal products. Therefore, any target for an arbitrary frugal product will usually need to be tightly followed with very little room for slack. In other words, the *actual value* (A) of a measurement for a target is given by,

$$A = T + E \tag{9.2}$$

where, T is *true value* of the quantity/property being measured and E is *measurement error* with E accounting for random or chance variations observed in actual measurement of *T*. Therefore, any activity of frugal products should have the lowest *E* possible since this will directly influence the goal of reaching the target. In other words, measurement bloc of Fig. 9.1 should have the lowest *E* even when the average of the target is same as *T*. Hence purveyors have to consider and coordinate all *E*s coming from all activities—involved in realizing a product—to design effective quality control systems for satisfying customers. Moreover, the complex nature of frugal products entail use of experimental tests that hew themselves to the necessary levels of standards required for gaging quality in frugality (Rao 2017b). Such standards should be set in terms of the high accuracy and high precision required of the procedures, tests and sensors involved in confirming output of frugal products for quality.

The quality control chart developed by Shewhart (1980), a schematic shown in Fig. 9.2a, is an important *classical* tool for informing when a corrective action is required. The chart in Fig. 9.2a shows three horizontal lines denoting upper and lower control limits and nominal or average value (Juran et al. 1974). The horizontal lines control for chance variation wherein numerous small random causes would contribute to the spread of the frugal product-characteristic-statistic around its average or nominal value. Any value falling outside these limits points to a major deviation that needs to be addressed. From a frugal standpoint, the limits have to be set at smaller values to check for any drifts from nominal values and also maintain stringent tolerances for robust functionality of the product. Such tightness in control limits necessary for frugal products is qualitatively compared in Fig. 9.2a with control limits for traditional products. Also, the drift to another nominal/average value of a concerned process should be ideally avoided as depicted in Fig. 9.2b. Furthermore, control charts should be run continuously to reap their benefit of monitoring a process for any slightest drift that would be crucial to quality from a frugal viewpoint. It should be noted that automation should be used where it is necessary to achieve tight tolerances of the control chart to avoid human error. The employment of automation in any aspect of frugal-product-production should consider the tradeoff between human error and necessity to maintain low cost.

An existing technique called six sigma popular at the industrial behemoth GE® should be adopted by makers of frugal products. This is because the six sigma tolerates very few defects by maintaining very tight standard deviations from the mean. In so doing the yield of a process is improved drastically while forcing the output from the concerned process to hew very close to the mean. The tighter control necessary for frugal products makes the six sigma technique a natural choice.

The *Pareto* principle (Juran 1975) popularly applied in *classical* quality problems, should be revisited for frugal products. This is because not just the few as expounded by the *Pareto* principle but the entire gamut of aspects involved in production of frugal products might be important for success. This is because control of factors not included in the 20% group might contribute little improvements that could cumulatively aid in achieving the total quality of the frugal product.

The importance of research and development (R&D) has been emphasized by Rao (2017a) for frugal innovations in general. The R&D activity will contribute immensely to every activity leading to the successful debut of frugal products. In

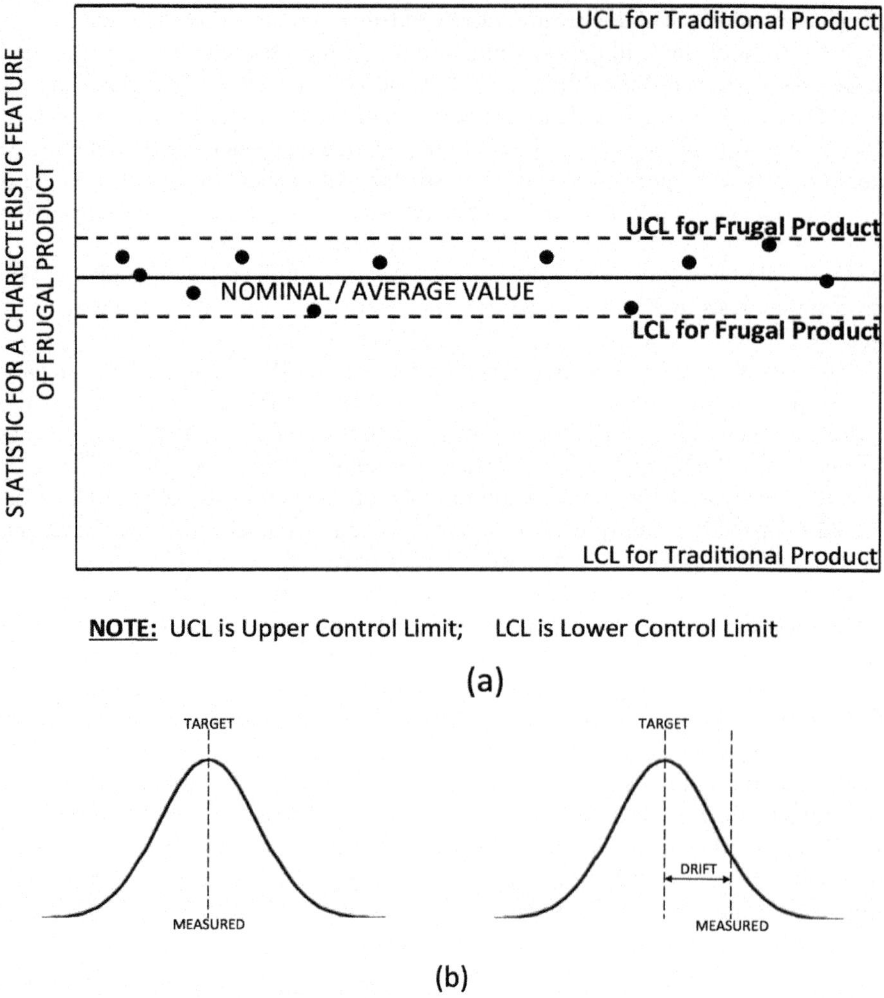

Fig. 9.2 Shewhart control chart (**a**) and unnecessary drift in process-performance (**b**) for a frugal product. Adapted from (Juran et al. 1974)

this regard, R&D efforts should be utilized to delve into techniques such as six sigma and others for tightly controlling the production of frugal products. Besides, proper R&D efforts can in the first place help in honing in on quality problems in a product thereby avoiding incorrect identification of trouble spots.

9.3.1 Quality Audits and Policy

Quality assurance is an important feature of any business entity and is more so with organizations dealing in frugal products. The high costs typically associated with quality vetting (Juran et al. 1974) are an antithesis to undertaking of frugality by an entity. These costs could be significant for smaller entities or job-shops vis-à -vis larger entities capable of mass production. However these costs are worth bearing by any serious activity, including business, in frugal innovations due to the balancing act required in designing products from scarce resources while aiming for excellent functionality. Therefore, quality control should be the most significant activity undertaken for any serious forays into frugality. Such control can be made official by including it in an existing quality audit department or creating one from scratch. Such a department can coordinate efforts including pilot studies to confirm the high levels of quality needed for frugal products before they reach the market. An important aspect of quality for frugal products is convincing the customer about the quality credentials of a frugal product despite its low price. Here again, a quality-department can make a difference by working with marketing staff in beaming the superior quality of frugal products at low cost.

A quality audit department for frugal products would differ from its traditional brethren in ideally closing the quality gap (Juran et al. 1974) in the product development process. The quality gap comprising understanding, design, process, operation and perception gaps (Juran et al. 1974) helps in honing onto and subsequent control of areas affecting quality. In the frugal case, the quality gap between customer expectation and product creator can be ideally closed by involving all sections of an organization in the design process so as to inform the entire workforce about rigor associated and expected of frugal products. In fact, the same quality control principles that apply to frugal products also apply to the internal workings of the makers of these products. Hence, everybody from sales and administration to design and production should be groomed on the importance of quality to products built out of scarcer resources. Moreover, such a department should also keep an active log or history of events that lead to failures which later generations of workers can use to advantage for ironing out quality issues for frugal products. The effectiveness of any quality control program for frugal products depends directly on how well quality is institutionalized by the company or entity involved.

Companies and entities connected with frugal products should make detailed quality manuals that shed light on the importance of frugality and why rigor is necessary in making these products successful. These manuals will summarize all quality control charts mapping the tight tolerances required of products and the ensuing rigor needed in various departments of the company in achieving it.

The purveyors of frugal products should also hew to environmental and other regulations set at global, state and national levels. These rules should be used by a maker to achieve/improve the functionality of their products built out of scarce resources. Moreover, makers of frugal products should also utilize *life cycle analysis* (LCA) in the initial stages of FDA to confirm the energy credentials of the product.

The audit department should also check results of LCA and include in its quality-assessments both accuracy of LCA and the product specifications. Instilling LCA into design and quality control will truly make the product a sustainable tour de force.

In case of frugal products and frugal services, the quality control should be an ongoing process that does not cease to function. Continuous monitoring for quality throughout the lifetime of any entity involved in making frugal products is a must. In fact, the emphasis on rigor, on a continuous basis, in any activity related to creation of frugal products should be enshrined in the quality mission statement (Juran et al. 1974) of the maker. Also continuous benchmarking should be sought against a frugal product's own stringent features in addition to competitors' winning aspects. In particular, internal benchmarking will also be significant for frugal products vis-à-vis their traditional brethren.

The traditional constituents of mission statements related to markets, technology, competition, performance etc. (Juran et al. 1974) should also include frugal-related information where necessary. Overall, mission statements should be drafted carefully so as to give goals that can be achieved by various sections of a company for realizing frugality within various constraints of the maker or company. In fact the quality mission for frugal products should be so pervasive as to make individual workers aware of the stringent norms expected of them. This will enable a coordinated effort rooted in rigor and also hewing stringently to quality control for making frugal products eventually succeed in the real world.

Unlike quality control for traditional applications, development of frugal products requires coordination of all involved, i.e., upper, middle and lower ranks of the workforce. In fact, even supply chains, if any are involved, should be made quality conscious from a frugal perspective. This is because the frugal nature necessiates tight control of quality wherein even the most insignificant issues at any level should be addressed effectively. Also, employees at all levels of the organization should be encouraged to submit their ideas—possibly based on their insitu observations—for improving quality many of which could facilitate successful quality control. The involvement of the entire workforce in being attentive to even the minutest details should be encouraged considering the constraints and also rigor native to frugal products.

Lastly, companies should at least give thought to the tradeoffs between stringent quality control and customer-needs. Such tradeoffs can help in reining in costs by focusing on quality that is commensurate to the needs of the customer. In other words, the rigor in control of quality can be relaxed and, accompanying costs lowered, provided the functionality satisfies customers over the life of the frugal product.

9.4 An Example

This section presents the quality control of a frugal shaft made of steel, whose design details have been reported by Rao (2017b). The selection of this simple example is based on the current non-existence of real-time frugal applications and also any detailed treatment is beyond the scope of this chapter.

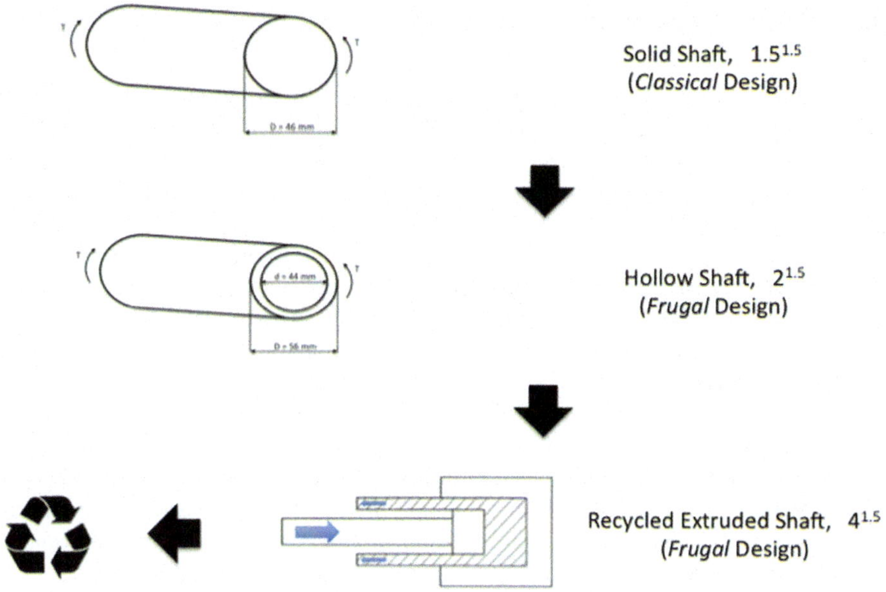

Fig. 9.3 Design of a frugal shaft (Rao 2017b)

Figure 9.3 shows a shaft designed through the *frugal design approach* (Rao 2017b) with an *outside diameter* (OD) of 56 mm and an *inside diameter* (ID) of 44 mm. Such a shaft has been produced with just enough steel so as to attain the diameters shown in Fig. 9.3 through the *indirect extrusion* process. Therefore effective *quality control* would require extrusion dies to possess high precision in terms of both surface texture and tolerances that is commensurate to achieving in turn the proper texture and tolerances on these diameters. In other words, die surfaces should possess roughness values much less than 0.8 μm to achieve tolerances as tight as ±0.025 mm.

Although, these are crude approximations based on existing processes, the frugal shaft needs such high precision in the processes used for its production with defect rates ideally close to zero. Quality control for such tight tolerances on the frugal shaft would neccessiate tighter control limits and also avoidance of any drifting, see Fig. 9.2a, b. In fact, such high precision required of tooling has several ramifications including for instance the selection of a proper extrusion machine-tool able to precisely align its constituent parts and also withstand higher loads stemming from the precision required in the shaft. Lastly, *six sigma* principles should be adopted in the quality control of this shaft to keep both defects and process-variations at a very low level. This would aid achievement of narrow tolerances while not wasting material.

9.5 Benefits of Quality Control

The combination of frugal design and quality control will improve the usability of frugal products. The verification of quality will also lend credence to the usage of the factor of frugality, a scientific tool of ecodesign, for quantifying frugality crucial to many sustainable solutions. The meticulous adoption of rigorous design and quality control during and also after product-realization would plausibly unleash large quantities of such products from various sectors into mainstream economy for widespread sustainable development. Other than environmental impact, stringent quality control of frugal products and/or frugal solutions will also go a long way in successfully realizing the goal of achieving robust functionality at lower costs using meager resources.

The problem of "chronic-waste" (Juran et al. 1974) will actually be more important to frugal products. The term refers to waste in resources of a company in trying to firefight products, after releasing into market, with shoddy quality. The makers of frugal products will realize their margins only if they tightly control the quality of all aspects of their business in realizing their products. Therefore stringent quality control is more important from the angle of chronic waste for frugal products.

Effective quality control also significantly helps in understanding the needs of customers (Juran et al. 1974). The process of quality control actually helps gage the fitness of use from the customer's perspective. In particular, focusing quality control on the needs perceived by the customer of a frugal product could significantly aid in plausibly avoiding rigor and associated costs with some features of the product. Such customer focused needs can be effectively realized by using the scientific principles of FDA to unleash quality frugal products in various sectors. Also, mass production of frugal products designed through FDA is possible provided proper stringent quality checks are in place to strictly rein in costs by avoiding outlays associated with fighting quality problems.

Moreover, discrepancies in quality of frugal products should lead to introspection by creators resulting in newer features and products. Since frugal innovations have appeared only recently, quality control will be crucial to the frugal product development process since deficiencies in the initial wave of products would hold lessons for futuristic products. Moreover, although effective quality control helps stick to specifications, they can also help R&D activities of the purveyor in achieving breakthroughs when fixing a quality-deficiency, e.g. away from control limits, that leads to a new version or a completely new type of frugal product.

In summary, this chapter stresses on the need for an important aspect of new-product-development called quality control, which though popular with traditional products and services carry serious connotation for frugal solutions. The road to success will be paved only through institutionalizing quality control in all aspects of producing frugal products. Accordingly, this chapter has stressed on the need, in the first place, to follow systematic design procedures such as the *frugal design approach* for methodical creation of frugal products from scratch. Second, frugal products designed rigorously should be subject to stringent quality control for both customer satisfaction and reining in of costs. In doing so, this chapter has extended some of the concepts of traditional quality control to the domain of frugal products. A rule of thumb would be to make traditional quality control tighter for all processes/activities, technical and non-technical, associated with the making of frugal products.

Questions

The questions to this chapter assume that the reader is well versed with the subject of quality control and related statistical analyses. If not, the motivated reader is encouraged to complete the undergraduate course on the same. These are also sample problems showing the significance of quality control to *frugal engineering*. Quality control is very important for confirming the *frugal* credential of parts/products while also aiding design itself for lowering costs.

1. Explain the importance of tolerances to various features of *frugal engineering*. In doing so, explain its significance to all the pillars of *frugal engineering* including design, manufacturing and others.
2. Inspection for tolerances includes expensive equipment like *coordinate measuring machine* (CMM). Justify the use of such equipment which are very important and indispensable to *frugal manufacturing*.
3. The *Taguchi loss function* (TLF) or *loss cost* is given by,

$$\mathrm{TLF} = \frac{\mathrm{RC}}{(\mathrm{LSL} - \mathrm{T})_2}[(\mathrm{Y} - \mathrm{T})^2 + \sigma^2]$$

   where RC is replacement cost; LSL is lower specification limit; Y is mean of quantity observed during manufacturing; T is target value of quantity from design; and σ is standard deviation of quantity from manufacturing. Explain the significance of the TLF to *frugal engineering*.
4. Is the closer setting of control limits depicted in Fig. 9.2 arbitrary. In other words, can we set these limits as tight as one wants to. Explain your answer using the concepts of *statistical process control* (SPC).
5. In continuation of the previous question, explain how *frugal manufacturing* influences the setting of control limits on the Shewhart control chart. Also, explain the reverse influence of Shewhart control chart on *frugal manufacturing*.
6. Company A has decided to follow the principles of *frugal engineering* for the mass production of its parts. As part of quality control it is using *acceptance sampling* wherein inspection is limited to random samples taken from the population of mass produced parts. However, stringent quality control in *frugal engineering* requires 100% inspection which will increase costs. What should company A do according to the tenets of *frugal engineering*?
7. Inspection techniques are very important to products realized with *frugal engineering*. Explain whether either or both of destructive and non-destructive techniques are important to *frugal engineering*.

8. Quality control concepts were propounded by pioneers such as Deming, Juran and Taguchi. One of the above questions also briefly pointed the reader to *Taguchi's loss function*. Explain which of the concepts established by these pioneers would be relevant to *frugal engineering*. The reader should read these different quality control concepts and pick relevant approaches for *frugal engineering*.

References

Agarwal N, Grottke M, Mishra S, Brem A (2017) A systematic literature review of constraint-based innovations: state of the art and future perspectives. Trans Eng Manag 64(1):1–13

Blanton G (1998) Hidden costs to society. Qual Dig 18(6):18

ISO (1994) ISO 9000 handbook, 2nd edn. Irwin Professional Publishing, Fairfax, VA, USA

Juran J (1975) The non-pareto principle; mea culpa. Qual Prog May:8–9

Juran J, Gryna F, Bingham R (1974) Juran's quality control handbook. McGraw-Hill, New York, USA

Knorringa P, Pesa I, Leliveld A, van Beers C (2016) Frugal innovation and development: aides or adversaries? Eur J Dev Res 28(2):143–153

Prabhu J (2017) Frugal innovation: doing more with less for more. Philos Trans R Soc A: Math Phys Eng Sci A375:20160372

Rao B (2013) How disruptive is frugal? Technol Soc 35:65–73

Rao B (2017) Advances in science and technology through frugality. IEEE Eng Manag Rev 45(1):32–38

Rao B (2017) Revisiting classical design in engineering from a perspective of frugality. Heliyon 3(5):e00299

Rao B (2019) The science underlying frugal innovations should not be frugal. R Soc Open Sci 6(5)

Shewhart W (1980) Economic control of quality of manufactured product. American Society for Quality Control (ASQC), Milwaukee, WI, USA

Tiwari R, Bergmann S (2018) What pathways lead to frugal innovation? Some insights on modes & routines of frugal, technical inventions based on an analysis of patent data in German auto components industry. Working Paper, No. 105, Hamburg University of Technology (TUHH), Institute for Technology and Innovation Management (TIM), Hamburg

Tiwari R, Herstatt C (2012) Open global innovation networks as enablers of frugal innovation: propositions based on evidence from India. Hamburg, Germany. Institute for Technology and Innovation Management Working Paper, Hamburg, Germany

TQM (1997) A Manifesto of TQM (1-4)—quest for a respectable organization presence. Societas Qualitas

Zeschky M, Widenmayer B, Gassmann O (2011) Frugal innovation in emerging markets: the case of Mettler Toledo. Res-Technol Manag 54(4):38–45

Chapter 10
Combating the Covid-19 Pandemic with Advanced Frugal Innovations

10.1 Introduction

The ongoing COVID-19 pandemic is a turning point in our history that has influenced our existence on several fronts. Other than the direct and dangerous impact on health, COVID-19 has also adversely affected the global economy. It has even changed the way we work and live due to our cognizance of the ensuing threat. COVID-19 has significantly battered the idea of globalization–but globalization will survive due to its allure not least, from a *frugal* perspective. *Frugal engineering* needs *globalization* to obtain cost-and-quality benefits by trading in a world without borders! Even without a *frugal* mindset, the cost-and resource savings together with typical reductions in associated emissions would continue endearing *globalization* to society at large.

Advanced frugal innovations started making their appearance early on in the pandemic and continue to do so as we are going through the third excruciating year. This chapter has been written to make the reader aware of the potential and healing-power of *advanced frugal innovations* during the COVID-19 pandemic and its applicability to other foreseeable crises.

10.2 Tests, Ventilators and Vaccines!

The COVID-19 pandemic has seen *advanced frugal innovations* coming to the rescue of the masses in various countries. A separate section in Chap. 4 highlighted some of the mobile-phony-based AFIs that sprung up to combat the pandemic. As mentioned in Chap. 4, these AFIs using *smartphones* and *apps* showed superior performance in diagnostic tests for COVID-19. Some of the AFIs used *smartphones* for affordable microscopy-to detect Covid–and also disease-epidemiology. Other than these diagnostic tools, several designs and, in some cases, actual mechanical ventilators were realized as AFIs. In fact, Chap. 4 covered one such ventilator *OneBreath*, developed

© The Author(s), under exclusive license to Springer Nature Singapore Pte Ltd. 2024
B. C. Rao, *Frugal Engineering*, Design Science and Innovation,
https://doi.org/10.1007/978-981-99-9700-8_10

by a *Stanford University* researcher, for resource poor regions that can also be utilized for emergency services such as a pandemic in rich and poor settings alike. Another brilliant *advanced frugal* realization of a ventilator was achieved by the teaming of Ford Motor Company and 3 M wherein their design takes contaminated air and pulls it through a 3 M filter and subsequently pushes it to the sealed mask using an air blower sourced from the Ford F150's front seat assembly (Popular Mechanics, 2020). A sketch of this improvisation for a ventilator attempted by Ford and 3 M is shown in Fig. 10.1. Yet another ventilator whose design was rooted in frugality came from the *University of Minnesota's Earl E. Bakken Medical Devices Center.* This design, with emergency and/or resource-poor applications in mind, came out of the need of a simple device that is easier to build affordably in such situations. Thus, the resulting ventilator made out of a simple crank slider mechanism is shown in Fig. 10.2.

Along with these implements for fighting the pandemic, vaccine, the weapon nipping COVID-19 in the bud, is also an AFI from the perspective of this book. An advanced workforce involving talented researchers from universities and companies have been working on the underlying science for years that bore fruit in the last couple of years. Costs were curtailed in avoiding animal-based tests (Ritskes-Hoitinga, 2022) due to the urgency of the situation. This decision was taken with due scientific rigor since there were troves of past animal data that were tapped into to cut time and costs. This is a hallmark of AFIs where quality or rigor is not sacrificed even under emergency situations and science, many times at cutting-edge, is utilized to arrive at the best possible solution. The quality of functionality was excellent as is evident in

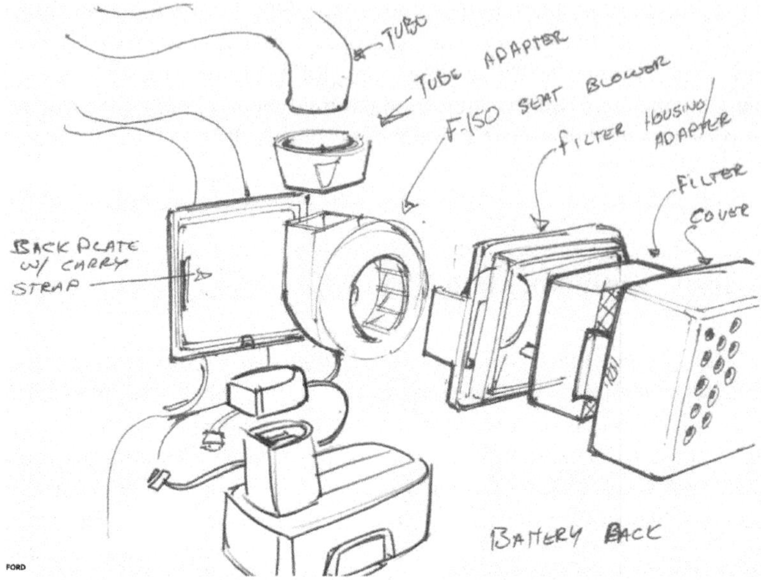

Fig. 10.1 Ventilator design using 3 M filter and Ford F150's air blower (Popular Mechanics, 2020)

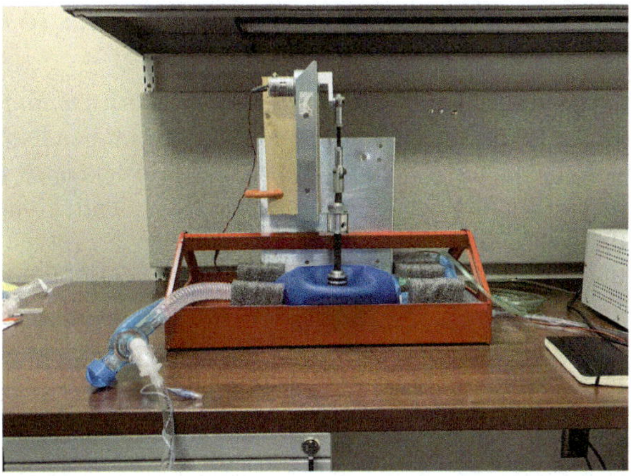

Fig. 10.2 A simple ventilator created using a crank and slider mechanism (https://cse.umn.edu/me/news/medical-devices-center-creates-new-ventilator)

the timely cure that saved countless lives. Although arrival of vaccines was somewhat hastened by the somber nature of the situation, it was advances in science together with *advanced frugal* settings that drastically reduced their time of development and creation. *Advanced frugal* settings refers to artificial restraints on resources, which during the pandemic was pursued due to paucity of time and also availability of both scientific resources such as supercomputers and knowledge, models, data, etc. All of this aided faster response in giving quality jabs and thereby reducing the span of the pandemic.

Generally, most of the solutions realized during the pandemic had some or a lot of features characteristic of *advanced frugal innovations.* In other words, most of these efforts were directed at keeping costs low; assembling from off the shelf parts/products; and aiming for high quality. All of these features arose naturally without anyone thinking about AFIs per se. This is because quality was the foremost requirement due to a push to stem mortality and also suppress infections from spreading due to the very contagious nature of the disease. The cost has to be kept low to support the fast diffusion of cures through all the strata of societies world over. Such fast diffusion is dictated by the uncertain and also dangerous nature of any pandemic! Other than these tell tale characteristics of AFIs, another hallmark of these solutions was their speedy arrival or fast response (just like for vaccines) in this emergency scenario. It can be argued that the typical simple designs of AFIs aids in their speedier development and faster fabrication. Moreover, a lot of the efforts in containing COVID-19 involved *do it yourself* (DIY) techniques that involved both novelty and improvisation on the go. A combination of frugality-based principles and speed of completion, due to frugality, aided significantly in getting many of these AFIs to the market.

10.2.1 3D Printing and Smart Phones to the Rescue

A special section has been devoted to *additive manufacturing* (AM) or *3D Printing* and mobile phony due to their use in several pandemic related AFIs. This book has described many examples of *smart phones* and *apps* going into the making of AFIs, many of which were used during the pandemic. *Smart phones* were ubiquitous in the war against COVID-19 since they have evolved into "computers-in-a-pocket" that are capable of many functionalities. *Smart phones* were used in diagnostic tests and surveys to assess infections at a human and group level respectively. This in addition to their use as a communication-tool that aided quicker dispersion of global-information to keep people informed about the many aspects of the disease.

3D Printing or *Additive manufacturing* (AM), described in the previous chapter, was put to widespread use in this pandemic either in making the complete product or its components. The bottlenecks in supply chains, of necessary clinical products, due to lockdowns were mitigated to a great extent by onsite printing of such products. Cost-effective nasal-swabs, indispensable for testing purposes, were *3D-Printed*, see Fig. 10.3, in some countries using *FormLabs*' printers (Decker et al., 2021). Several AFIs or their components utilizing *3D Printing* have been presented in Chap. 4. Among these AFIs, SPOT developed at UIUC was noteworthy due to its ability to give really fast COVID diagnostic results in about 30 min. SPOT was made affordable, just ~$90, through a portable device many of whose components were *3D Printed.* 3D *Printing* is a versatile manufacturing process that works on both metallic and non-metallic materials. The proliferation of affordable *computer aided design* (CAD) tools, over the years, has significantly increased the ability of even amateurs to design and subsequently input the resulting STL files to *3D Print* all kinds of products/components. *Additive manufacturing* helped both amateurs, using DIY techniques, and professionals, in major corporations/public entities, to try out various possibilities, during the pandemic, in less time. AM together with other technologies aided quick testing and manufacturing of several alternatives for various COVID-19 related clinical products.

Both these technologies enabled advances in science through frugality being put to good use during the pandemic for creating successful solutions. Such timely concerted action from confluence of several advanced technologies would not have been possible even about a couple of decades ago. A weaker form of *globalization* together with banding-together-of-technologies significantly aided in the relatively short and successful fight against COVID-19.

10.3 Abetting Spread of Innovations and not the Pandemic

As mentioned at the beginning of this section, *globalization* helped in the dispersal of AFIs around the world. Numerous ingredients of *globalization* made this fast and effective spread possible. Although global supply chains took a beating, their

Fig. 10.3 *3D printing* of nasal swabs for COVID-19 diagnostics (Decker et al., 2021)

frame work in terms of communication, transport, and infrastructure, to name a few, provided ease of sharing pandemic-related knowledge, products and, services. The sheer amount of knowledge that became available from formal and informal sources was immensely helpful globally for countless workers/researchers for attacking the virus. The progress in technology including Internet, mobile phony and *3-D printing* were indispensable in the fight against COVID-19.

A world rendered borderless due to *globalization* and technology helped in the fast definition of the problem and also fast solutions through AFIs. Speed is of essence during any emergency situation on the scale of a pandemic and the world is in a better position than when similar catastrophes were inflicted in the past.

10.4 The Indian Story

India was also active in combing for solutions and this task was rendered challenging by the country's size and its population. The resulting solutions were mostly of the *advanced frugal* type necessitated by stringent quality requirements that was ingeniously pulled off by brilliant researchers/workers and a society that is cost-conscious.

India started early on in the pandemic by inter alia looking for its own vaccine. This foray into COVID-19 research was facilitated by the existence of world-class biotechnology hubs in India with many focusing on medicine. Such initiatives have resulted in *Bharat Biotech International Limited* developing and also manufacturing Covaxin™, India's own very successful vaccine for the pandemic. *Bharat Biotech* developed this in collaboration with India's *National Institute of Virology*. The other major Indian company that responded in a timely manner was the *Serum Institute of India*, whose R&D and rich experience in mass-production of vaccines in general was immensely helpful while rolling out the Indian version of *AstraZeneca*'s vaccine (*Covishield*) that it developed with the *University of Oxford*, UK. Other than vaccine, Chap. 4 has described an AFI, by an Indian researcher, using *smartphones* along with CRISPR gene-editing technology and *machine learning* to detect viral infection. The faster response to the pandemic was an imperative that was not lost out on the world. This was also true of India and its exposure to COVID-19 brought out the best in talent, policy-making, leadership and all other ingredients necessary for a fast and effective response through AFIs.

In all of this, the *frugal* expertise of India was very useful in combating the pandemic. In particular, Indian researchers and relevant workers, through their ingenuity and dint of hard work, worked round the clock against the pressures of the pandemic in realizing the cures. The hard work is noteworthy because the talented workers responsible for the vaccine worked almost 24 X 7 for weeks, without visiting their homes, after the pandemic's onset in India.

10.5 Frugal Innovations are Here to Stay!

The chapters of this book have taken the reader through many aspects of *frugal innovations* in general and, *advanced frugal innovations* in particular. The astute reader would have noticed the significant potential of these innovations/products, which were brought to bear upon the COVID-19 pandemic. The improvisation-on-the-go and other frugality-based-methods are not mundane because the pandemic needed the science and its advances for the research and subsequent development of novel cures at a pace never seen before. And AFIs resulted from these fast scientific endeavors because the underlying theme of such innovations supports rapid response in emergencies. The many AFIs presented in this book including the ones for the pandemic are a testimony to the potential of *frugal innovations*. These examples in myriad sectors make it clear beyond doubt that *advanced frugal innovations* are here to stay to help humankind fight against emergencies and other crises.

References

Decker SJ, Goldstein TA, Ford JM, Teng MN, Pugliese RS, Berry GJ, Pettengill M, Silbert S, Hazelton TR, Wilson JW, Shine K, Wang ZX, Hutchinson M, Castagnaro J, Bloom OE, Breining DA, Goldsmith BM, Sinnott JT, O'Donnell DG, Crawford JM, Lockwood CJ, Kim K (2021) 3-dimensional printed alternative to the standard synthetic flocked nasopharyngeal swabs used for coronavirus disease 2019 testing. Clin Infect Dis 73(9):e3027–e3032. https://doi.org/10.1093/cid/ciaa1366

Popular Mechanics (2020) Ford is using seat ventilation fans to build respirators

Ritskes-Hoitinga M (2022) World view-medical regulators: look beyond animal tests. Nature 604:599

Chapter 11
Frugal Computing for Artificial Intelligence and Other Applications

11.1 Introduction

A summary, to buttress the reader's understanding, of the content covered until this chapter is presented in this paragraph followed by the need for computing as an AFI. As mentioned in earlier chapters, *climate change* (CC), demographics and other natural crises (Rockström et al, 2009) together with human made ones such as the global financial crisis of 2007–09 are goading us to embrace *sustainable development* (SD) in all endeavours. These crises could be collectively referred to under *climate change.* Other than capping energy consumption, constraints also have to be imposed on resource consumption to help in mitigation and adaption efforts against these crises. These large-scale constraints accordingly require major shifts in the use and application of *science & technology* (S&T) underpinning all the products and services used globally. In this regard, the advent of *advanced frugal innovations* (AFIs) is a welcome development that has warranted modifications to engineering itself for coping with *climate change* (Rao 2019, 2021). *Advanced frugal innovations* have been defined in this book as low-cost products scientifically engineered out of minimal resources that give the best functionality without sacrificing on safety (Rao, 2017a). Therefore, building AFIs from scratch, or from retrofitting existing products, needs a gearing of engineering to make it majorly dependent on constraints on resources and costs. Accordingly, *frugal engineering* (FE) was defined as design and fabrication of parts/products for best functionality achieved with minimal resources at low-cost (Rao, 2022). AFIs have been observed in nearly every sector of industrial activity from healthcare, automotive and particle-physics to astronomy (Rao 2013, 2017a). In fact, the theme of an AFI as an affordable and sustainable alternative to traditional products/services for good functionality underlies most of the applications that have appeared over the past couple of decades and several of which have been documented in this book. Moreover, AFIs typically need cutting edge *research & development* (R&D) activities for their successful realisation (Rao, 2017a). And this trend will continue since *climate change* has serious ramifications for our existence

© The Author(s), under exclusive license to Springer Nature Singapore Pte Ltd. 2024
B. C. Rao, *Frugal Engineering*, Design Science and Innovation,
https://doi.org/10.1007/978-981-99-9700-8_11

in this world and policymakers, the world over, should consider development of AFIs in various sectors (Liefner et al., 2020) including computing and services.

Computing is the backbone of economic activity whose need will only grow with time due to not least, increased adoption of *artificial intelligence* (AI). However, ever increasing computing activities with their attendant emissions will also typically increase the carbon footprint. This chapter presents the framework for a *frugal* version of computing called *frugal computing*. *Frugal computing*, an *advanced frugal innovation*, is made possible by using lesser resources and maintaining low-cost while achieving best performance. Resources here refers to the data, software, hardware and other paraphernalia going into a computing application.

The previous century saw rapid progress in communications, internet, software, e-services and other digital tools. A crucial element underpinning these developments is computation that was and continues to be a backbone of all manner of e-tools and digital applications. However, computing needs power and hence contributes to *climate change*. Computing is a complex activity involving the algorithm, and resulting software, along with the hardware necessary for the execution to obtain necessary results. With ever increasing demand for computing from various sectors, there is a crucial need to make computing "frugal" for reaching maximum efficiency with minimal resources and hence low-cost. Again, the term resources as used in this chapter includes data; software; power-consumption; hardware; labor; and other paraphernalia needed to accomplish a computing task. Therefore, f*rugal computing* (FC) is an AFI because it achieves best performance with minimal resources at low-cost!

This chapter presents the concept of an *advanced frugal* version of computing called *frugal computing* (FC) with emphasis on different aspects of computing. Other than presenting the underlying concept, this work also gives some possible methodologies for achieving frugality in computing. This chapter is very brief and introductory requesting interested researchers to work and modify these approaches for effective FC. The concepts here are rudimentary without details but they point to many open-ended problems in FC specially when considering AI applications including *machine learning*.

11.2 Frugal Computing

Computing is the backbone of a vibrant economy and high on the list of *greenhouse gas* (GHG) emitters (Nicole Jones, 2018). An extreme example of such emissions is the massive amounts of energy consumed in blockchain computations since the advent of cryptocurrencies (Nicole Jones, 2018). The coming of *artificial intelligence* (AI) is also putting a strain on computing resources and the associated carbon footprint. Examples in AI including *large language models* (LLMs) and other *machine learning* (ML) tools consume large amounts of energy, not least due to their widespread adoption. Hence there is a need for an *advanced frugal* way of computing, called *frugal computing* (FC), that conserves on resources including fuel consumption and, hence emissions and, at the same time maintains quality functionality at low-cost. Any algorithm/software/hardware (or their combinations) based on these features would qualify to being an *advanced frugal innovation.*

Therefore, FC as an AFI will need to be frugal in resource-and-power consumption while giving best performance at low-cost. There are different ways of achieving such frugality in computing, or *frugal computing*, and some of these are as follows:

a. ***Super-efficient algorithms***: An obvious way to achieve FC would be to make the algorithm(s) for a given application efficient. Algorithms, the first stage in any computing activity, should be created with a focus on using lesser resources in terms of data and effective use of hardware. Algorithms can also be designed to be adaptive to dynamically account for the changing environment of a computing activity.
b. ***Unsupervised learning***: Unsupervised or unconditional learning needs fewer but "powerful" models for extraction of problem features to get better generalisation. Therefore, lesser dependence on data sets and hence plausibly lesser resources! However, the limited dependence on data should be built into the learning model as a constraint or a penalty for large data sets.
c. ***Use Training Data or Data Sparingly***: Use of very large data sets having millions of points is not uncommon in *machine learning* applications. Hence use of "portions" of such large data sets or even sequential/incremental approaches using one data point at a time for computational purposes should be encouraged. However, care should be exercised to avoid *overfitting* where smaller training sets in *machine learning* have been known to overfit polynomial models.
d. ***Database of What Works***: Since there is always an efficient computing solution for a given application, a database could be created (or an existing one used or modified for better usage) and maintained that keeps track of such best algorithms/models in various applications. Subsequently, ML (or other computing tools) could be used to make optimal use of such a database in arriving at a FC solution for a new application.
e. **Complex systems**: FC should make use of complex systems/complex networks wherein multitudes of simple agents through their interactions affect emergent behaviour that could be beneficial to a given application (Cook, 1998). The application domain of complex systems theory is large (Grisogono, 2006) with a recent effort having applied it to *frugal grid electric vehicles* (Rao, 2020). In computing, a software could be the agent that could communicate with other softwares to reach an emergent behaviour that leads to FC.
f. **Reinforcement learning/neural networks**: Reinforcement-learning and other AI tools could be created for incentivizing *frugal* behaviour while penalizing excesses. Such hardwiring of frugality into *machine learning* could go a long way in establishing FC on a firmer foundation.
g. ***Smart Hardware***: Hardware is the other significant pillar of FC that needs to be addressed in the future. The hardware itself can be made smart to gage the load and accordingly take action to minimize carbon footprint. Hardware comprises the basic chips to all manner of computer-architectures/designs that when designed

for frugality can help computing maintain a low carbon footprint for even high-emission applications.

11.3 Some Ramifications and Conclusion!

Frugality in computing would not only help save on resources and, hence costs, but also have a plausibly positive influence on the time of computing. Generally smaller datasets/training sets; smaller but effective codes/models; and effective use of databases would together or individually lessen the time of computing.

The ever-increasing global adoption of computing will increase resource-consumption and hence the carbon footprint. Therefore, FC needs to be adopted globally on a very large scale to realise the significant savings in resources, with attendant reductions in the carbon footprint and, costs. In fact, ideally, every computing-need needs to be *frugal* to relieve the strains put on our resources and time. This becomes even more important due to computing's ubiquity. In other words, FC should become *de rigueur* in the expanding world of computing.

Many of the methodologies suggested in Sect. 11.2 for realising *frugal computing* can be implemented by adopting existing tools. However, some of the methodologies suggested in Sect. 11.2 are currently not available (but hopefully possible and not a figment of my imagination!) and should be the focus of research. A hybrid approach combining some or all of the methodologies of Sect. 11.2 should be considered in making computing *frugal.*

Lastly, government-bodies and policymakers should understand the gravity of computing's influence on *climate change* and the significant potential of FC whose leverage can make computing climate friendly. Moreover, FC can significantly improve the efficiency of computing's use in understanding *climate change* not least, by realistic simulations.

Climate change is revamping human endeavors for aiding humanity's mitigation and adaptation to a warming world. Computing, an indispensable activity to any economy, is being increasingly used in the expanding realm of *artificial intelligence.* Accordingly, a *frugal* version of computing called *frugal computing*, presented and discussed in this chapter, would be significant against this backdrop. *Frugal computing*, an instance of *advanced frugal innovation*, consumes minimal resources while aiming for best affordable performance. Some rudimentary methods to implement and bring out the efficacy and cost-effectiveness of *frugal computing* have been briefly discussed.

References

Cook RI (1998) How complex systems fail. Cognitive Technologies Laboratory, University of Chicago, Chicago IL, pp 64–118

Grisogono A (2006) The implications of complex adaptive systems theory for c2. Technical report, Defence Science and Technology Organisation (DSTO), Australia

Jones N (2018) The information factories. Nature 561:163–166

Liefner I, Losacker S, Rao BC (2020) Scale up advanced frugal design principles. Nat Sustain 3:772

Rao BC (2013) How disruptive is frugal? Technol Soc 35(1):65–73

Rao BC (2017) Advances in science and technology through frugality. IEEE Eng Manage Rev 45(1):32–38

Rao BC (2021) Frugal manufacturing in smart factories for widespread sustainable development. R Soc Open Sci 8:210375

Rao BC (2019) The science underlying frugal innovations should not be frugal. R Soc Open Sci 6(5)

Rao BC (2020) On complex systems of adaptive frugal products. R Soc Open Sci 7(7)

Rao BC (2022) Frugal engineering. In: Régnier P, Frey D, Pierre S, Varghese K, Pascal (eds) Handbook of innovation and appropriate technologies for international development. Edward Elgar Publishing Ltd, UK

Rockström J, Steffen W, Noone K, Persson Å, Chapin III FS, Lambin E, Lenton TM, Scheffer M, Folke C, Schellnhuber H, Nykvist B, De Wit CA, Hughes T, van der Leeuw S, Rodhe H, Sörlin S, Snyder PK, Costanza R, Svedin U, Falkenmark M, Karlberg L, Corell RW, Fabry VJ, Hansen J, Walker B, Liverman D, Richardson K, Crutzen P, Foley J (2009) Planetary boundaries: exploring the safe operating space for humanity. Ecol Soc 14(2):32

Chapter 12
Combating Poverty with Frugal Innovations

12.1 Introduction

According to past estimates by researchers at the World Bank, more than 2 billion people are exposed to penury—living under \$ 2-a-day—in the developing world. Although *globalization* has uplifted millions of people trapped in poverty around the developing world, other supplementary measures to alleviate poverty over the long term are needed to cover the vast numbers of the poor. At the turn of the century, natural threats, such as *climate change*, have made *sustainable development*, with its ensuing *innovations*, the basis for the futuristic betterment of humankind. Consequently, the engineering of *frugal innovations* introduced and described in the previous chapters is a potent tool for fighting the maladies of our society. This chapter attempts a stab at one such malady, poverty from the perspective of *sustainable development* against the backdrop of globalization using *frugal innovations*. This chapter shows the flexibility of *frugal engineering* in tackling poverty while catering to the advancements in *science and technology* for better environment- and people-friendly products. Accordingly, this chapter presents a new framework for tackling poverty using *frugal engineering* of *frugal innovations*.

Globalization has facilitated the growth of several Asian and South American economies, including the BRIC *(Brazil, Russia, India and China)* countries, and thereby catapulted them to the status of a rising star (World Bank[1], 2010). Consequently, the globalization-induced economic betterment together with lower-income inequality has reduced the proportion of the destitute from 43% in 1990 to 21% in 2010 based on a poverty line of \$1.25-a-day (Ravallion, 2013). Although the current estimate for poverty's definition is 1.90 per day (Osendarp et al., 2022), the message of this chapter is not lost as it attempts to outline a solution for the alleviation of poverty. China alone has lifted about 662 million people since 1981 (Chen & Ravallion, 2010) due to globalization-led growth, agrarian reforms and lower taxes albeit accompanied by higher- income inequality (Ravallion & Chen, 2007). It should be noted that though globalization seems to be out of favor currently—such pessimism

© The Author(s), under exclusive license to Springer Nature Singapore Pte Ltd. 2024
B. C. Rao, *Frugal Engineering*, Design Science and Innovation,
https://doi.org/10.1007/978-981-99-9700-8_12

being brought about by the pandemic and current geopolitics—and has lost some of its steam, this chapter supports globalization and strongly believes that globalization will continue to shape our future. Although the global poverty rate has diminished by half in the past two decades, freeing the remaining poor from the clutches of extreme poverty is a challenge not least due to their location in fragile areas such as Congo and Somalia (The Economist[1], 2013). According to new estimates for 2008 by the World Bank (Chen & Ravallion, 2010), 801 million people in developing countries were exposed to extreme poverty, i.e., $1-a-day, and 2.47 billion people lived under $2-a-day. In contrast, nearly 3 billion people were employed globally in 2011 according to the *International Labor Organization* (ILO, 2012). However, according to the same report (ILO, 2012), the *employment-to-population* ratio has been, on average, decreasing thereby pointing to rising unemployment. Moreover, in 2011, the share of working-extreme-poverty ($1.25-a-day), working poverty ($2-a-day) and vulnerable employment, i.e., no job security, of the total global employment was 14.8%, 30% and 49%, respectively (ILO, 2012). Additionally, the financial crisis has dealt a severe blow to the ranks of the employed and thereby added to the total number of poor globally (ILO, 2012). This total number is bound to increase further as the world population is forecasted to touch the 8 billion mark *circa* 2030 (PRB[1], 2012). Therefore, a large section of society, comprising the poor on the one hand and the unemployed on the other, has to be sustained.

The next front in the fight against poverty will also have to grapple with planetary threats, i.e., global warming, dwindling resources and others (IEA, 2009; MA, 2005; IPCC, 2001; Rockström et al., 2009), facing humanity from here onwards. The ensuing wave of innovations for combating these threats and promoting *sustainable development* is a force for good that should be enlisted in any new effort to alleviate poverty.

The poverty alleviation schemes proposed until now and described in the next section do not account for the collective action of factors apparent above, i.e., natural threats, innovation and globalization. This effort presents a framework for fighting poverty when viewed through a prism of these three factors. It is to be noted that the proposed framework is aptly suited to a democratic government possessing good rule of law, policies and intellectual property rights. Before proceeding any further, this chapter believes in referring to countries other than the developed ones as developing. The economies of the erstwhile third world have been spurred into action since the advent of globalization with both cost-effective technological advances and structural reforms. This is slowly bridging the gaps between them and their rich counterparts.

12.2 Traditional Causes of Poverty

Some of the important factors influencing poverty include under-nourishment, economic inequality, infrastructure, credit and insurance market, institutional (or government) failure, misguided government policies, democracy, literacy, environment and the management of commons (erosion of local commons) (Dasgupta, 1998).

However, in the foreseeable future, population explosion will by far have the greatest impact on poverty due to its influence on many of the above-mentioned factors.

The traditional schemes, reported to date, to fight poverty include monetary aid (Burnside & Dollar, 2000), philanthropy (Reiser, 2008), financial tools (Bossone and Sarr, 2002; Moser, 2006; Smillie, 2009), innovations (Dutz, 2007a, b) and other approaches and policies (Banerjee and Duflo, 2011).

12.3 Need for the Triumvirate

Planetary crises, including climate change, water scarcity, loss in biodiversity and dwindling resources have elicited responses that range from leading a spartan lifestyle without innovations (McKibben, 2010) to an all-out embrace of "green technologies" (Brand, 2009). The latter alternative affords a unique opportunity for mounting a new attack on poverty. The advancements in various disciplines, including *science and technology* (S&T), should be harnessed to innovate in all aspects of the human endeavor for combating planetary crises while also aiming for inclusive growth that uplifts the poor. In other words, principles of sustainability—to conserve earth's resources and also mitigate the impacts of planetary crises—should be adhered to while innovating to improve the standards of living for the poor and well-off alike. In particular, *sustainable* innovations would provide employment opportunities besides creating myriads of useful products and services for raising standards of living. And the focus on sustainability would lead to innovations, i.e., products and services, made from minimal amounts of raw materials and other resources drawn from the environment. The *frugal* nature of these innovations would typically lower their costs thereby facilitating deployment in poorer regions. Moreover, the positive aspects of globalization, such as the relatively free flow of goods, services and knowledge (Das, 2010), should be utilized to facilitate the widespread advent and diffusion of these *sustainable innovations* or *frugal innovations*. Therefore, the trio of factors, i.e., natural threats, innovation and globalization, form a potent mix for any poverty alleviation solution(s) from this century onwards. The ordering of these factors also hints at planetary threats spurring *frugal innovations* beneficial to the environment, whose advent and diffusion are abetted by the gears of globalizaton. Henceforth, this trio of factors will be referred to as the "triumvirate" with the proposed framework being rooted in it.

12.4 Central Theme of the Proposed Framework

Economic growth harnessing the triumvirate would unleash *sustainable development* in myriad sectors of an economy thereby spawning environment-friendly products and services. This growth will be stretched over the long term, not least due to the nature of natural threats, to gainfully employ people in a *sustainable* economy and also raise their living standards, with the *frugal* goods and services created in that economy. Ergo, *sustainable development* through *frugal innovations* is the overarching theme of the proposed framework with the key components described below.

Sustainable Development: The term *sustainable development* (SD) was coined in 1987 by the *United Nation's World Commission on Environment and Development* and defined as "development that meets the needs of the present without compromising the ability of future generations to meet their own needs" (WCED, 1987). SD has been acknowledged for its critical role in poverty alleviation by the *Convention on Biological Diversity* (CBD, 2006).

The use of SD from this century onwards is a must as humankind strives to enjoy the benefits of economic growth against the backdrop of dwindling resources, natural threats and man-made crises. This is because factors such as shortage of resources, urbanization, population explosion, *climate change* and other calamities beg for solutions having a minimal impact on the environment and also conserving earth's resources while developing technologies & infrastructure for civilian use or rendering service as a public good or even making policies. Therefore, any venture—tangible or intangible, scientific or unscientific, novel or incremental, for civilian or military use (Jorgenson et al., 2010)—should be based on the concepts of SD. Furthermore, the concepts and benefits of SD should "trickle" down through the economic ladder to the lower-income and poorer segments of society, whose larger numbers justify imbuing products and services realized for their consumption with sustainability.

As for the actual penetration of SD, to date, both developed and developing nations have been striving to create new technologies and infrastructure for SD. The geoengineering schemes (Nature[2], 2012) and development of alternate-fuel systems (Nature[3], 2012) are but a few prominent examples of this tangible application of SD. In this regard, Chap. 4's examples bring out the various *advanced frugal innovations* in a wide range of sectors as building blocks for SD. AFIs are rooted in sustainability and as such should be used as basic LEGOesque blocks for building our sustainable world. As for applications in the intangible domain, monetary aid should be coupled with SD–*see tenet 1 in Sect. 5*. Another intangible outcome would be the empowerment of the *International Monetary Fund* (IMF), or another global body, for the oversight of activities in SD, the world over. However, such a decision should be based on the unanimous consent of developed and developing countries alike with the IMF, in particular, needing a makeover for having drawn flak in recent years (Stiglitz, 2002).

Frugal Innovations: In addition to being a pool of cost-effective talented workforce, developing economies, epitomized by India, have also pioneered the process of innovating under conditions of scarcity (Srinivas and Sutz, 2008) as also pointed out in the chapter on *grassroots frugal innovations*. The advent of *frugal innovations* (FI) resulted in adapting technologies for making low-budget vehicles and innovative healthcare for the masses of the Indian sub-continent as pointed out in this book and also in (Knowledge@Wharton, 2009; The Economist[2], 2009; Novy-Hildesley, 2010).

Products such as bamboo microscopes (http://www.dare.co.in/people/featured-innovation/a-microscope-made-from-bamboo.htm) and improvised vehicles used by Indian farmers (The Economist[3], 2009) have been the products of this newfound ingenuity and thrift. Beyond ordinary products, companies are applying the concept of FI in the high-technology sector by coming up with products such as hand-held electrocardiograms. Therefore, the economic usage of resources to innovate frugally fits in well with the concept of sustainability.

Eventually, public awareness toward frugality should go beyond conserving resources to maintaining healthier lifestyles to avoid for instance obesity which is endemic to the developed world. Besides combating cardiovascular disease, type 2 diabetes and some cancers (Bleich et al., 2008), successful regulation of obesity would have a positive impact on, among other factors, the conservation of both resources and energy (Jacobson & McLay, 2006). The concept of frugality has a lot in store for developed countries as well. They should take cues from their developing brethren for learning to innovate and live under *frugal* conditions which would enable the developed world to weather crises in the future.

12.5 Poverty Fighting Mechanisms Rooted in Sustainability

This section presents various schemes for engendering SD through FI for alleviating poverty that is still prevalent around the world. These schemes should be utilized to lay the ground for more FIs in the foreseeable future. Consequently, where not mentioned explicitly, for brevity FI is also implied in the schemes presented below whenever a reference to SD is made.

1. Continuation of Monetary Aid: There is some evidence that monetary aid influences growth in countries with good institutions and good policies (Burnside & Dollar, 2000). Therefore, aid to developing countries should be contingent on their enshrining the concepts of SD in their policies to facilitate growth for combating poverty. In this regard, the plan presented by Glenn Hubbard (Hubbard & Duggan, 2009) should be tweaked to direct aid at SD with the private sector as one of the players. With respect to this tenet, initiatives such as the *climate fund* (Nature[4], 2009), initially proposed by the UK prime minister Gordon Brown, with US $ 100 billion donated annually by wealthy nations for spurring green activities in developing countries is a laudable step in the right direction. Another noteworthy scheme to raise funds for SD in developing countries through taxation has been proposed by Barbier (Nature[5], 2012). In contrast, poverty alleviation efforts, such as the Conditional Cash Transfer (CCT) (The Economist[5], 2010) and Bill Gates' plan at a G-20 summit (Nature[6], 2011), do not involve activities rooted in SD. This is because a significant portion of monetary aid to poor countries is utilized to combat pressing issues such as hunger and disease. However, the adoption of SD would gradually lessen these maladies of our time since SD also entails better living standards.

However, rich nations like the OECD countries have to live up to their pledges on monetary aid toward developing ones and they have been found wanting in stepping up their commitments (OECD[4], 2010) which is due to the financial crises (Nature[6], 2011) and also pandemic related expenditure including stimuli. In addition, rich countries should also mull over debt relief to many developing nations. Here again debt relief should be made contingent on involvement in SD. Such relief could provide a respite and a plausible motivation to start afresh in SD.

2. Fraternity of Nations: With natural threats and man-made crises looming large in this era of *globalization*, it is but natural for the developed and developing nations to work together to seek solutions to the global challenges of this century. The concept of intergovernmental collaboration through multilateralism propagated by Keynes (Ruggie, 1982) to abet balance-of-payments equilibrium in individual economies should be aptly extended for achieving concerted collective action among various nations for all-round SD. The fraternity of nations would seek cost-effective and efficient solutions to numerous global challenges, including large-scale projects such as geoengineering, where any individual nation might find it difficult or impossible to go it alone. The fraternity should also focus on topics in macroeconomics, and microeconomics, having an impact on SD, such as exchange rates and remittances, whose market-based mechanisms should be monitored and ably regulated to, in turn, ease the cross-border flow of products, services and knowledge. Besides, these synergies could be used to shore up the potential GDP trend rate of rich nations by revolutionizing their innovation processes with the frugality of emerging markets. But other than sporadic efforts through monetary aid (previous tenet) and a few schemes (Nature[7], 2009; The Economist[6], 2009), intergovernmental interactions have been weak not least, due to weaker health of the global economy due to current geopolitics, pandemic and also the lingering effect of the financial crisis.

3. Involvement of the Private Sector: Other than government, the private sector is also a cornerstone of SD, whose involvement is critical to the successful penetration of sustainability in our lives. While governments lubricate the free markets with both national policies and infrastructure conducive to SD, the private sector and other players should utilize this milieu to innovate frugally, where possible, for developing technologies, products and services. The private sector is usually marked by the following features: corporate focus on tasks for creating shareholder value; survival of the fittest; promoting competition; productivity improvement; performance and, last but not least, fostering innovations to exploit better business opportunities (Netter & Megginson, 2001; Nidumolu et al., 2009). Therefore, the profit-making activities of the private sector should be leveraged for SD thereby fighting poverty. Consequently, the work by Nidumolu and co-workers (2009) encourages companies to embrace FI targeted toward low-earners, comprising "bottom of the pyramid", to face the eventuality of this century and beyond. It also motivates companies to be the first in their herd to adopt SD to their advantage. In this regard, witness the successful efforts of Unilever whose *Sustainability Living Plan* has been very effective over the years in profiting from frugally innovative products for these masses (The Economist[7], 2012). Such a spurt in *frugal innovations* by multinationals, especially Unilever,

Procter & Gamble and GE, also aids effective diffusion of SD into poorer regions and provides poor people with cheap and environmentally benign products and services. Other companies of the developed and developing world should take cues from these innovation-leaders and thereby contribute to the fight against poverty. Even venture capitalists and other entrepreneurs scouting for suitable start-ups should plow their money into SD-related technology and services at the earliest. To date, venture capitalists have had their investments plowed into life sciences, biotechnology (Nature Biotechnology, 2012), space, electric vehicles, etc., that are deemed to have "market value". However, the success of Unilever and GE attest to the rewards for catering frugally to the masses.

Among areas where the private sector is falling short of progress is the use of financial instruments for spurring green activities. Such instruments would go a long way in supplementing traditional cap-and-trade schemes for stemming *Green greenhouse gas* (GHG) emissions. Apropos, Rao and co-workers (2008) have developed the concept of a financial derivative linked to innovations for roping in the free markets to invest in green technologies. Exchanges such as the *Chicago Mercantile Exchange* (CME), a pioneer in derivatives trading, should design such an *innovation derivative* for involving the free-market system in SD. The *Small and Medium size Enterprises* (SME) is another area whose innovative output is critical to SD. The work by Beck and co-workers (Beck et al., 2005) has shown a positive association between the size of the SME sector and GDP per capita growth. SMEs which typically cater to niche areas, and more so to the middle- and lower-income segments, should be encouraged to adopt sustainability-based practices to frugally innovate for the masses. Lastly, philanthropic activities of the private sector, such as *Corporate Social Responsibility* (CSR) (Heal, 2005), *Social-Enterprise* (DTI, 2003) and *For-Profit Philanthropy*, should be associated with SD thereby having a positive impact on social welfare.

4. A Tiered Workforce for the World Economy: In recent years, technology-savvy workers from China and India have been instrumental in offering the Western economies skilled labor at relatively lower costs. In fact, India has seen an increase in the number of science and technology graduates from 1.8 million in 1980 to 8 million in 2004 (Krishna and Chandra, 2009). Although this huge workforce is mostly effective for the run-of-the-mill activities outsourced from rich nations, China and India, have also been developing resources for a knowledge economy, comprising facilities and a competent workforce, in science, engineering and management (Freeman, 2005). This fact is attested by the presence of excellent institutions like the *Indian Institutes of Technology* (IIT), *Indian Institute of Science* (IISc) and the *Indian Institutes of Management* (IIM), to name a few.

Other than using revenues drawn from this labor force—to typically supplement a diminishing workforce in OECD countries (Nature[8], 2010; PRB[2], 2009)—for poverty alleviation in emerging markets, this vast workforce, including the knowledge workers, could be agents of change bringing about widespread SD. Their ingenuity in creating FIs for companies, such as GE's research center in India, is a testimony to their usefulness in the global spread of SD. Therefore, this workforce should be categorized into three tiers to facilitate contributions to SD commensurate to the abilities of individual workers. The diving up of labor into skill-based tiers will improve the efficacy of remunerating people who like doing what they do. In the process, the tier system could increase the ranks of the employed, thus influencing the upliftment of the downtrodden. Consequently, a physicist involved in critical research, i.e., tier #1, would not find the work content of routine tasks carried out in call centers or other technical setups commensurate to his abilities. In so doing, the creativity and talent, whether in a scientific discipline or otherwise, innate to a worker plus his desire would put him in a commensurate tier. Although technocrats are indispensable for helping workers to innovate effectively (Bhidé, 2008), they should not exceed the numbers of workers, at various skill levels, who actually carry out the innovation process. A critical mass of workers should be maintained in all the tiers who innovate, frugally where possible, for progressive SD under the auspices of skilled managers. The three tiers of the workforce are:

a. ***Tier 1***: The global economy should make good use of workers with advanced educational degrees and/or superior talents available in rich and developing nations for innovating in crucial scientific and non-scientific disciplines. Some examples of such workers include scientists, mathematicians and economists, to name a few, who are eminent in their fields. Therefore, a partnership between rich and developing countries for tapping into the pool of tier 1 workers would go a long way in devising *frugal* products and services in critical sectors such as healthcare and aviation and also attacking problems of global concern, especially natural threats.

   The ongoing polymath project at the *Massachusetts Institute of Technology* (MIT) is an example of global partnership in open science for solving difficult problems in mathematics (Nature[9], 2009). Other examples include MIT Media Lab (http://www.media.mit.edu/) tapping into Indian tier 1 workers and the *Bill and Melinda Gates Foundation* fighting infectious diseases, among other issues, afflicting the developing world (http://www.gatesfoundation.org/global-health/pages/overview.aspx) through research carried out by groups involving tier 1 workers from emerging economies. Governments around the world are realizing the significance of "brain circulation", i.e., facilitating cross-border movement of elite workers, for solving complex problems such as *climate change* (Nature[10], 2012). Their actions in terms of immigration policies and infrastructure will enhance the mobility of tier 1 workers in the foreseeable future.
b. ***Tier 2***: The tier 2 group of workers would include the next level of skilled workers who currently comprise the backbone of all the commercial and non-commercial activities both scientific and non-scientific including outsourcing.
c. ***Tier 3***: The tier 3 class of workers includes the bulk of the masses in the developing markets. Workers in this tier would be associated with sectors such as agriculture and forestry and hence in the immediate vicinity of mother nature. Their thrifty solutions to problems and knowledge of natural habitats would contribute to FI globally. Such workers should be empowered and encouraged to stay put in these positions since their work and ideas have a significant contribution to devising

frugal solutions that can help mitigate *climate change* and other crises of our time. However, there is room for climbing up to tiers 1 and 2 as individuals aspire to work in those tiers and continue to progress in both skill levels and education.

5. Relevant Sectors for Local Growth and Employment: Barnett (2004) has stressed the need for directing innovations to issues relevant to the needs of a developing country without imitating the creativity of other nations, particularly developed ones. There is a significant number of consumers in China and India, whose numerous wants present a rare opportunity to harness widespread innovations for their betterment. Here the recent financial crises have lessons for the developing world to avoid being a predominantly export-oriented economy and cater more to their domestic consumers partly to escape unscathed from the global ramifications of crises in the future. The specific needs of developing countries, which could be commercial and/or humanitarian, could be addressed through the rigors of SD. In this regard, the advent of FI in China and India has faciliated the creation of several low-cost goods and services for the low-income group. This concept has also been exported to the developed world to cater to their cost-conscious consumers. By consuming minimal resources for their realization, FI has become an indispensable tool for SD impacting both low-income people and cost-conscious higher-income populace. But developing countries should also temper the use of technological automation, and therefore, boost employment numbers, in their quest to sustainably satisfy the needs of multitudes of consumers.

Agriculture is a critical sector in developing countries employing the bulk of the poor in rural areas (FAO & World Bank, 2000). The significance of this sector will increase in the coming years with a burgeoning population. The work by Sanchez (Nature[11], 2009) comments on the need to empower the farming sector in developing countries with good agricultural know-how for tackling the problem of world hunger effectively. Such a burst in populace warrants the increase in yield of agricultural products for feeding society at large. Although natural farming practices such as organic farming should be encouraged for SD, technological advances in farming, such as drip irrigation (Brinegar & Ward, 2009) and *genetically modified Organisms* (GMO) should be mulled upon with research efforts targeted at studying the benefits of the latter which has drawn flak in recent years (Nature[12], 2010; OECD[5], 2000). A report from the *National Research Council* (NRC) in the US, while referring to the risks of transgenic crops has also highlighted the advantages of some genetically modified crops such as reduced insecticide use, higher yields, minimum tillage and lower production costs (NRC, 2010).

In contrast, a recent effort (Nature[13], 2012) has also reported that improved irrigation techniques and optimal nutrient supply are paving the way to *sustainable agriculture* with yields that could be increased up to 70%. However, agricultural advances, used individually or in combination (Nature[14], 2011), contribute to SD through their high yield from optimal use of resources. Agriculture, after forestry, has also shown its potential for mitigating GHGs by sequestering carbon and other elements through the usage of croplands and grasslands (EPA, 2010). In fact, a study has shown a reduction in GHGs through land saved due to a highly productive agricultural system employing modern technology (Burney et al., 2010). The use of agriculture for producing bio-fuels that mitigate GHGs in the long run is being researched for supplanting fossil fuels (Vuichard et al., 2009).

Infrastructure, including roads, railways, etc., is another sector that needs to be upgraded in several developing countries to facilitate economic growth. Infrastructure development has been shown to have a statistically positive impact on the economic growth of China (Démurger et al., 2002). The development of "green" infrastructure is an important exercise in SD whose presence would also allow the fostering of other sustainability-based technologies and services. The development of green infrastructure would boost economic growth, improve employment numbers and concomitantly aid the fight against poverty while being benign to the planet. An apposite development is China's sustainability-based plans for managing water, a scarce resource, involving relevant infrastructure to aid in conserving this resource in the future (Nature[15], 2011).

Banking and financial services, important pillars of any economy, would have new roles, including fostering innovations through innovation derivatives reported previously as economies adopt SD. They have to encourage, inter-alia, the poor to save (Collins et al., 2009) and aid easy access to banking and credit by the poor. Microfinance, through microcredit, has been shown to encourage entrepreneurship among the poor studied in India (Banerjee et al., 2009) and the Philippines (Karlan & Zinman, 2009). However, unlike the political backlash against microfinance in Bangladesh and India (The Economist[8], 2012; The Economist[9], 2010), a form of microfinance in Africa using pooled savings has recently improved the lives of about 4.6 million poor people in 54 countries of Africa, Asia and Latin America (The Economist[10], 2011). Especially, Indonesian microfinance banks have been profitably giving credit to the poor (The Economist[11], 2011). Malpractice by microfinance entities notwithstanding, the governments of developing countries should encourage microfinance and put suitable regulations to foster its growth to encourage the poor to innovate frugally for widespread SD.

Beyond traditional sectors, developing countries should also embrace other issues imperative from the perspective of SD. For example, India has to combat the impending water scarcity due to both groundwater depletion (Kerr, 2009) and the receding of glaciers through *climate change*. Developing, and also developed, countries should harness market forces to reward forestation and their conservation (Streck et al., 2008, Nature[16], 2010) for mitigation of GHGs. China and India are also the destinations of ever-increasing amounts of global-electronic-waste from hardware, such as computers, whose lifetimes keep decreasing (Yu et al., 2010). The resulting pollution of the environment due to "informal" recycling of various electronic components begs for a solution rooted in the precepts of SD. A country such as South Africa should devote ample resources to combat AIDS. South Africa is home to 17% of the people infected with HIV (Nature[17], 2010) and economic growth through SD is possible only after the spread of AIDS is stemmed effectively.

6. Educational Reform: Universities, which are the hotbeds of both innovation and learning, are crucial to building a competent workforce for SD. The bulk of the workforce, as outlined in the 4th tenet, in developing countries needs to be trained for knowledge-based activities that are required for widespread SD. In this regard, China and India are witnessing the building of resources focusing on educating their foreseeable knowledge-based workforce (Nature[18], 2011). Success in training students to acquire a sustainable bent of mind relies on incorporating the following features. First, these educational programs should keep the curricula abreast of the latest developments in relevant planetary affairs (Nature[19], 2010). And second, such curricula should also be complemented by creativity and lateral thinking while avoiding rote learning.

Besides spurring creativity, educational reforms should also sow the seeds of frugality among the global population for achieving higher standards of living while both conserving resources and avoiding damage to the environment. As far as basic education is concerned the spirit of frugality together with creative thinking should be encouraged in the first twelve grades after kindergarten so that the spirit of abstemisim is ingrained in the formative years.

Universities would also be indispensable to carrying out research on sustainability-based concepts such as FI, which is yet to mature and hence still in an evolving status, for empowering SD. Accordingly, despite the sagging demand for PhD workers in a majority of OECD countries, China and India are producing and absorbing larger numbers of PhDs into their workforce (Nature[20], 2011). Such a tier 1 workforce should be utilized for both research and diffusion of FI into society at large. It is also critical to stress the quality of PhDs (Nature[20], 2011) and other tier 1 workers produced as opposed to their quantity, as is currently the trend in many countries around the world.

12.6 The Road Ahead

The method of randomized trials (Banerjee and Duflo, 2011) should be applied, in future work, to the individual schemes for lending credence to the frugality-based framework of this effort. While aiming for all-round SD the schemes of this chapter have been framed for society at large, including the poor at the grassroots of society. The poor are important stakeholders in SD—besides women and the elderly—who are close to nature; large in number and lead a frugal life. Frugality, the linchpin of this framework, would be the basis of synergistic relations with the poor to tap into their rich and significant knowledge and potential for FI.

Sustainable development over the long term is inevitable. The rising tide of SD would raise the economic boats of both developed and developing countries. Therefore these schemes could also be applied to developed countries where poverty headcounts, based on a 50% median income threshold, have risen in the past (OECD[3], 2008). This is because natural threats and the concept of sustainability are universal to this planet, and therefore, offer a unique opportunity for viewing poverty through a new perspective.

A sustainability-based framework has been presented in this chapter for tackling penury from this century onwards. This is because planetary threats, such as *climate change*, loss in biodiversity and scarcity of resources, offer an unusual opportunity for a spurt in economic growth over the long term through sustainability. *Sustainable development* and its representative, *frugal innovations*, lend themselves to the triumvirate, i.e., globalization, innovations and planetary crises, which have been missing in existing strategies to combat poverty. The proposed framework stresses the use of innovations carried out under frugal conditions for widespread *sustainable development*. The poverty-fighting framework presented comprises several schemes encompassing the roles of the private sector and technology, to name a few. Furthermore, numerous examples of *sustainable development* in a wide range of sectors have been scattered throughout this chapter. Other than emerging markets, these tenets could also be employed to minimize traces of poverty seen recently in developed countries.

References

Arnold E, Bell M (2001) Some new ideas about research for development. Dan Minist Foreign Aff Partn Lead Edge: Dan Vis Knowl

Banerjee A, Duflo E (2011) PoorEconomics: a radical rethinking of the way to fight global poverty. Perseus Books, United Kingdom

Banerjee A, Duflo E, Glennerster R, Kinnan C (2009) The miracle of microfinance? Evidence from a randomised evaluation. MIT Poverty Action Lab, Cambridge, Massachusetts

Barnett A (2004) Research to poverty reducing innovation. A Policy Brief from Sussex Research Associates Ltd., UK

Beck T, Demirguc-Kunt A, Levine R (2005) SMEs, growth, and poverty: cross-country evidence. J Econ Growth 10:199–229

Bhidé A (2008) The venturesome economy: how innovation sustains prosperity in a more connected world. Princeton University Press, New Jersey

Bleich S, Cutler DM, Murray CJ, Adams A (2008) Why is the developed world obese? Annu Rev Public Health 29:273–295

Bossone B, Saar A (2002) A new financial system for poverty reduction and growth. IMF Working Paper 02/178

Brand S (2009) Whole earth discipline: an ecopragmatist manifesto. Viking-Penguin, New York

Brinegar HR, Ward FA (2009) Basin impacts of irrigation water conservation policy. Ecol Econ 69(2):414–426

Burney JA, Davis SJ, Lobell DB (2010) Forthcoming. Greenhouse gas mitigation by agricultural intensification. In: Proceedings of the national academy of sciences

Burnside C, Dollar D (2000) Aid, policies and growth. Am Econ Rev 90(4):847–868

CBD (2006) Global biodiversity outlook 2. Secretariat of the Convention on Biological Diversity, Montreal.

Chen S, Ravallion M (2010) The developing world is poorer than we thought, but no less successful in the fight against poverty. Quart J Econ 125(4):1577–1625

Collins D, Morduch J, Rutherford S, Ruthven O (2009) Portfolios of the poor: how the world's poor live on $2 a day. Princeton University Press, New Jersey

Das DK (2010) Globalisation: past and present. Econ Aff 30(1):66–70

Dasgupta P (1998) The economics of poverty in poor countries. In: Development economics discussion paper DEDPS/9, the suntory and Toyota international centres for economics and related disciplines (STICERD). (Also Symposium organized by Scandinavian Journal of Economics, Jan. 1997).

Démurger S, Sachs JD, Woo WT, Bao SM, Chang GH, Mellinger AD (2002) Geography, economic policy, and regional development in China. Asian Econ Pap 1(1)

Department of Commerce (2008) Innovation measurement: tracking the state of innovation in the American economy. A Report to the Secretary of Commerce by The Advisory Committee on Measuring Innovation in the 21st Century Economy, Washington, DC. (http://www.innovationmetrics.gov)

Dietz S (2007) The impacts of climate change: perspectives from the stern review. Brown J World Aff 13–2:173–184

DTI (2003) A progress report on social enterprise: a strategy for success. Progress Report by the Department of Trade and Industry (DTI), UK

Dutz MA (2007) Unleashing India's innovation: toward sustainable and inclusive growth. The World Bank, Washington, DC

Dutz MA (2007) Unleashing India's innovation. The International Bank for Reconstruction and Development-the World Bank, Washington DC

EPA (2007) Sustainability research strategy. Environmental Protection Agency (EPA) Washington, DC

EPA (2010) Inventory of U.S. greenhouse gas emissions and sinks: 1990–2008. Environmental Protection Agency (EPA), Washington, DC

FAO and the World Bank (2000) Agricultural knowledge and information systems for rural development (AKIS/RD). Strategic Vision and Guiding Principles-Food and Agriculture Organization of the United Nations and the World Bank, Rome

Freeman RB (2005) Does globalization of the scientific/engineering workforce threaten U.S. economic leadership?. NBER Working Paper 11457

Goyal A (2010) Forthcoming. Information, direct access to farmers, and rural market performance in central India. Am Econ J: Appl Econ

Heal G (2005) Corporate social responsibility: an economic and financial framework. Geneva Pap Risk Insur-Issues Pract 30(3):387–409

Hubbard G, Duggan W (2009) The aid trap. Columbia University Press, New York

IEA (2009) World energy outlook. International Energy Agency, Paris

ILO (2012) Global employment trends: preventing a deeper jobs crisis. Employment Trends Team of the International Labor Organization, Geneva

IPCC (2001) Climate change 2001: the scientific basis. Contribution of working group I to the intergovernmental panel on climate change, Cambridge University Press, New York

Jacobson SH, McLay LA (2006) The economic impact of obesity on automobile fuel consumption. Eng Econ 51(4):307–323

Jorgenson AK, Clark B, Kentor J (2010) Militarization and the environment: a panel study of carbon dioxide emissions and the ecological footprints of nations, 1970–2000. Glob Environ Polit 10(1):7–29

Karlan D, Zinman J (2009) Expanding microenterprise credit access: using randomized supply decisions to estimate the impacts in Manila. Yale University Economic Growth Center Discussion Paper 976.

Kerr WR, Lincoln WF (2008) The supply side of innovation: H-1B visa reforms and US ethnic invention. Harvard Business School Working Paper 09–005.

Kerr RA (2009) Northern India's groundwater is going going going.... Science 325:798

Knowledge@Wharton. Renault-Nissan CEO Carlos Ghosn: 'Now Is the Time for the Electric Car': (http://knowledge.wharton.upenn.edu/article.cfm?articleid=2364)

Krishna VV, Chandra N (2009) Knowledge production and knowledge transfer: a study of two Indian institutes of technology. Asia Research InstituteWorking Paper 121

MA (2005) Ecosystems and human wellbeing. Island Press, Washington, DC, Millennium Assessment
McKibben B (2010) Eaarth: making a life on a tough new planet. Times Books, New York
Moser CON (2006) Asset-based approaches to poverty reduction in a globalized context: an introduction to asset accumulation policy and summary of workshop findings. The Brookings Institution: Global Economy and Development Working Paper 01.
NAS (2009) In national academy of sciences speech, president Obama announces major investments in research and education, encourages NAS members to think about new ways to engage young people in science. President Barack Obama's Speech to the National Academy of Sciences (NAS), Washington, DC
Nature Biotechnology (2012) Biotech's wellspring: a survey of the health of the private sector 30(5):395
Nature[10] (2012) Research funding: global challenges need global solutions 490:337–338
Nature[11] (2009) A smarter way to combat hunger 458:148–149
Nature[12] (2010) What it will take to feed the world. An interview with Marion Guillou 464:969.
Nature[13] (2012) News: global survey reveals routes to boosting crop yields 488
Nature[14] (2011) Preventing hunger: biotechnology is key 479:471–472
Nature[15] (2011) World view: China's water crisis needs more than words 470:307
Nature[16] (2010) A green investment. 457:144–146
Nature[17] (2010) AIDS research must link to local policy 463:733–734
Nature[18] (2011) World view: university cuts show science is far from saved 469:133
Nature[19] (2010) Climate class for business schools 466:30
Nature[2] (2012) News: dumping iron at sea does sink carbon 487.
Nature[20] (2011) News feature: The Ph.D. factory 472:276–279
Nature[3] (2012) Perspective: opportunities and challenges for a sustainable energy future 488:294–303
Nature[4] (2009) News-environment: the globe's green avenger 460:454–455
Nature[5] (2012) Comment: tax 'Societal Ill's to save the planet 483:30
Nature[6] (2011) Editorial: innovative vision. 479:149
Nature[7] (2009) Boost the developing world 457:958
Nature[8] (2010) Advances in development reverse fertility declines 460:741–743
Nature[9] (2009) Opinion: massively collaborative mathematics. 461(15):879–881
Netter JM, Megginson WL (2001) From state to market: a survey of empirical studies on privatization. J Econ Lit 39(2):321–389
Nidumolu R, Prahalad CK, Rangaswami MR (2009) Why sustainability is now the key driver of innovation. Harv Bus Rev 56–64
Novy-Hildesley J (2010) By the grace of invention: how individuals power gevelopment. Innovations 5(1):7–24
NRC (2010) The impact of genetically engineered crops on farm sustainability in the United States. The National Academies Press, Washington, DC
OECD[2] (2010) The OECD innovation strategy: getting a head start on tomorrow. OECD Publishing, Paris
OECD[3] (2008) Growing unequal?: income distribution and poverty in OECD countries. OECD Publishing, Paris
OECD[4] (2010) Donors' mixed aid performance for 2010 sparks concern. Report by the OECD's Development Assistance Committee (DAC)
OECD[5] (2000) Genetically modified foods, widening the debate an health and safety. In: The OECD Edinburgh Conference on the Scientific and Health Aspects of Genetically Modified Foods
Osendarp S, Verburg G, Bhutta Z, Black RE, Pee SD, Fabrizio C, Headey D, Heidkamp R, Laborde D, Ruel MT (2022) Act now before Ukraine war plunges millions into malnutrition. Nature 604:620–624
PRB[1] (2012) World population data sheet 2012. Population Reference Bureau, Washington, DC

PRB[2] (2009) Global aging and the demographic divide. Population Reference Bureau, Washington, DC

Rao B, Sundararajan A, Rao S (2008) Trading of innovations to combat globalization and climate change. Chall: Mag Econ Aff 51–5:70–84

Ravallion M, Chen S (2007) China's (Uneven) progress against poverty. J Develop Eco 82(1):1–42

Ravallion M (2013) How long will it take to lift one billion people out of poverty?. The World Bank Research Observer

Reiser DB (2008) For-profit philanthropy. Brooklyn Law Sch Leg Stud Res Pap 124

Rockström J, Steffen W, Noone K, Persson Å, Chapin FS, Lambin E, Lenton TM, Scheffer M, Folke C, Schellnhuber H, Nykvist B, De Wit CA, Hughes V, van der Leeuw S, Rodhe S, Sörlin PK, Snyder R, Costanza U, Svedin M, Falkenmark L, Karlberg RW, Corell VJ, Fabry J, Hansen B, Walker D, Liverman K, Richardson PC, Foley J (2009) Planetary boundaries: exploring the safe operating space for humanity. Ecol Soc 14(2):32

Ruggie JG (1982) International regimes, transactions, and change: embedded liberalism in the postwar economic system. Int Organ 36(2):379–415

Smillie I (2009) Freedom from want: the remarkable success story of BRAC, the global grassroots organization that's winning the fight against poverty. Kumarian Press, Virginia

Srinivas S, Sutz J (2008) Developing countries and innovation: searching for a new analytical approach. Technol Soc 30:129–140

Stangler D (2009) The economic future just happened. Ewing Marion Kauffman Foundation, Kansas city, MO

Stiglitz J (2002) Globalization and its discontents. Penguin Books Ltd., London

Streck C, Sullivan RO, Janson-Smith T, Tarasofsky RG (2008) Climate change and forests: emerging policy and market opportunities. Brookings Institution Press, Washington, DC

Tella RD, MacCulloch R (2008) Gross national happiness as an answer to the Easterlin paradox. J Dev Econ 86:22–42

The Economist[1] (2013) Briefing poverty: not always with us

The Economist[10] (2012) Small wonder: helping the poor to save

The Economist[11] (2011) Indonesian microfinance: rich pickings.

The Economist[2] (2009) Lessons from a frugal innovator

The Economist[3] (2009) Cost cutting in Asia: a snip at the price

The Economist[4] (2009) First break all the rules: the charms of frugal innovation

The Economist[5] (2010) Anti-poverty programmes: give the poor money

The Economist[6] (2009) Face value: the alternative choice

The Economist[7] (2012) Consumer goods: fighting for the next billion shoppers

The Economist[8] (2012) Grabbing grameen

The Economist[9] (2010) Microfinance in India: discredited

United Nations Secretariat (2012) World urbanization prospects: the 2011 revision executive summary, report by the population division of the United Nations secretariat. United Nations, New York

Vuichard N, Ciais P, Wolf A (2009) Soil carbon sequestration or biofuel production: new land-use opportunities for mitigating climate over abandoned Soviet farmlands. Environ Sci Technol 43:8678–8683

WCED (1987) Our common future. Oxford University Press, World Commission on Environment and Development, Oxford

World Bank[1] (2010) World development indicators. The World Bank, Washington, DC

Yu J, Williams E, Ju M, Yang Y (2010) Forecasting global generation of obsolete personal computers. Environ Sci Technol 44:3232–3237

Chapter 13
The Future

Frugal innovations have come a long way from their inconspicuous beginnings at the grassroots level of societies in emerging markets such as India and China. The *grassroots* innovations have transformed into *advanced frugal innovations* which are created through the systematic application of scientific principles and cutting-edge research. Other than emerging markets, these innovations have now entered into mainstream markets of even rich nations of the *Organization for Economic Cooperation and Development* (OECD). As shown in this book with numerous examples, AFIs hold promise for abetting *sustainable development* and this is more so when AFIs are created from scratch through the *frugal design approach* described in this book.

The *frugal* nature achieved through scientific sophistication in myriads of AFIs reported in this book shows that it is possible to extract minimal resources while uplifting living standards across the world. The researchers who have developed these innovations have displayed time and again the ingenuity required to tackle our current crises. In fact, the increasing awareness of planetary crises including *climate change*, loss of biodiversity and ecosystems, pollution and depletion of resources are goading humans toward a *frugal* and hence sustainable living. As a result, knowledge workers from sciences, economics, socials, management and others are increasingly looking at ways to positively impact our environment. This is resulting in all kinds of clever AFIs for enjoying a higher quality of living in the present while also bequeathing a beautiful planet to future generations.

Frugal innovations should not be set aside due to their *frugal* nature which has been misconstrued by many in the society at large as being "cheap". On the contrary, examples of AFIs such as, though not limited to, particle accelerators, electrocardiograms and missiles attest to the sophistication of "simple" *advanced frugal* products. Therefore, the *frugal* nature of these well-engineered innovations encapsulates the mantra of taking little from *mother nature* but giving a lot to society at large. The significance of this statement will ring true as humankind enters into a long period of the increasing menace posed by *climate change*.

© The Author(s), under exclusive license to Springer Nature Singapore Pte Ltd. 2024

B. C. Rao, *Frugal Engineering*, Design Science and Innovation,
https://doi.org/10.1007/978-981-99-9700-8_13

13.1 India, the Pioneer in AFIs

This book has been made possible due to my first-hand experience of brilliant Indian AFIs. Indian scientists and engineers at India's pioneering space agency, the *Indian Space Research Organisation* (ISRO), have again shown their brilliance in the latest landing of *Chandrayaan 3*, see Fig. 13.1. Brilliant *research and development* done by ISRO in collaboration with India's premium academic institutions including the *Indian Institutes of Technology* (IITs); the *Indian Institute of Science* (IISc); and others have resulted in this massive success. The *Chandrayaan* missions are AFIs that ultimately achieved their success in accomplishing an arduous task at a price tag that might floor their peers in other countries.

The ingenuity and resourcefulness underlying the brilliant success of ISRO is implicit in the definition of an AFI given in this book. In other words, the trio of resource-effectiveness; cost-effectiveness; and quality-functionality can be achieved simultaneously through a resourceful *R&D* that can function under stringent constraints on resources and cost while wringing the maximum out of rigorous scientific principles. Industrialized countries should accordingly take note of the Indian resourcefulness that has gone into achieving the three goals of *frugal engineering*. In fact, the *factor of frugality* approach described in this book is actually directed at facilitating the creation of such "proper" AFIs from scratch. In other words, this book has laid the foundation for the methodological creation of low-cost AFIs that con-

Fig. 13.1 India's Chandrayaan 3 lifting off for a brilliant success! Image taken from https://www.isro.gov.in/chandrayaan3gallery.html

sider safety and quality-functionality despite their creation out of minimal resources. Therefore, AFIs are different from innovations that are light weight. In fact, innovations that are only light weight and/or put paid to safety and other similar philosophies will be plain "frugal-washing"!

13.2 Climate Change and Other Crises

The world has witnessed temperature-shattering records every year in the recent past due to the gradual encroaching of planetary crises typified by *climate change* (IPCC 2001; IEA 2009). Other planetary crises, including population explosion and decline in biodiversity and ecosystems, are making it imperative to learn to adapt to a different world in the foreseeable future (Rockström et al. 2009). The ongoing pandemic, though weakening gradually to endemics, has awakened our people to a new threat that needs a rapid and quality response. Other than natural threats, human-made crises such as the financial meltdown of 2008–09 and the ongoing war in Ukraine are also goading humans to help their unfortunate brethren in fulfilling their materialistic needs with products and services that are both subsidized and possess world-class quality. All in all, affordability and quality can be provided in most of the natural and human-made crises by the *frugal innovation* movement!

Advanced frugal innovations are a potent tool in the fight against *climate change*. The ever-present focus on conserving resources in the *frugal design approach* always leads to the development of products that are apt against the crises of this century and beyond. A caveat to be noted here is the assumption of having adhered rigorously to all the tenets of the *frugal design approach* in designing and fabricating these products. Therefore, a keen focus is required on exhaustively searching (or researching) for as many options as possible to save on resources in all areas of product development.

13.2.1 Climate Change is an Energy and Resource Problem

The heading of this subsection is partly based on the significance of tackling the global energy problem for adapting to and also controlling *climate change*. The fact that *climate change* is deeply connected to energy was made clear by MacKay (2009) in his excellent book on this topic. Readers are encouraged to delve into the work by MacKay (2009) to realize the energy angle on *climate change*. Other than energy, resources are also an important component of the *climate change* problem and resource consumption can be judiciously controlled through *frugal engineering*.

The *frugal* way of carrying out engineering will lead to lesser resource consumption which generally will involve lesser energy consumption and lesser attendant emission of pollutants. But, *frugal engineering*, with its "green" credentials, also encourages engineers, scientists, policymakers and other relevant stakeholders for maximal adoption of renewable energy. It should be noted that *frugality* enables savings of resources that are finite and cannot be squandered. Such an approach lowers energy consumption but the goal of tackling *climate change* also requires independent adoption of renewable energy. In other words, complete substitution of fossil fuels by renewable energy hypothetically will negate emissions feeding into *climate change*. Currently, these emissions are from human endeavors that are not all *frugal*. However, converting all engineering activities to *frugal* and complete dependence on renewable energy will be the best scenario for saving our planet. The availability of renewable energy does not give us the license to waste, as resources are finite!

13.3 Support from Governments

Against the above backdrop, suitable policymaking and federally-mandated regulations could greatly help in the uptake and diffusion of *frugal innovations* globally (Liefner et al. 2020). *Frugal innovations* of *grassroots* and *advanced* types need to be adopted on a very large scale considering the irreversibility of the planetary crises we are facing today. Even though the concept of *frugal innovations* took shape about a decade ago, the need to economically consume earth's resources has always been a theme that the world believed in. However, ideally, such a theme should have been adopted in tandem with the onset of *industrial revolution* in the seventeenth century. Even though second-guessing is inevitable in hindsight, a *frugal* attitude rooted in *sustainability* is not completely in vogue even today. Yes, governments around the world are waking up to this dangerous reality with countries such as India, China, Japan and several Western economies even having started battening down the hatches against *climate change*. But a lot more needs to be done federally and a gentle nudge, which some decades ago would have been salutary, should transform into coercion through national mandates.

Governments the world over should facilitate the ready adoption of scientific tools such as the *frugal design approach*. In particular, governmental and relevant non-profit entities should set up subsidized programs to empower designers, engineers and the like with *frugal* design through academicians, scientists and other qualified workers. Public entities should shoulder the main responsibility of creating awareness and also informing people in general about relevant institutions, workers and any other amenity for seeking design solutions rooted in frugality. Obviously, these services including actual design work should be offered free of cost to *grassroots* innovators. The private sector should also be encouraged to select and support, in terms of funding and knowledge, design work for promising *advanced frugal innovations* that possess the potential for both uplifting living standards of society at large and commercial success. Companies should also bankroll *grassroots* innovations which can have multifold benefits other than the primary one of uplifting the poor. Other than using *grassroots frugal innovations* for other applications in richer settings, their workings could even help the creation/modification of AFIs. Overall, this work urges the involvement of public, private and other relevant entities in educating, funding and empowering *frugal innovations* irrespective of whether they are *grassroots* or *advanced*.

13.4 More Research

Although this book has attempted to unfold the interesting and useful features of *advanced frugal innovations*, it is up to scientists, engineers, any other knowledge worker or even society in general for taking interest in these innovations and their significant potential to help save a world gradually changing for the worse. The pace of *climate change* and other crises will increase at a rapid clip if left unchecked. Other than planetary concerns, sophisticated *frugal* products can be a game changer when it comes to being a market leader for new products with their associated profits. Such profits can be accrued by companies and other entities while selling these low-priced and sophisticated products to uplift the living standards of society at large through their quality performance.

Among most of the activities affecting frugality, not least for lowering costs, design and engineering, being primary, should be carried out without cutting any slack. This is because *frugal design and engineering* (FDE) infuses the product with frugality from scratch. Secondary activities such as supply chain logistics, marketing and finance, to name a few, are important in lowering costs for products both *frugal* and the ordinary. In other words, secondary activities could lower costs even for bulky and bad quality products without focusing on the frugality necessary at the most basic level of product development. The secondary activities are also significant for lowering costs but only after FDE, and even these should be targeted only on *frugal* products. Hence, large-scale *research & development* activities should be undertaken to bring out products against rigorous constraints that need massive involvement of research culture. This is because existing AFIs and those yet to come would need cutting-edge research for their realization through FDE.

13.4.1 Modeling Challenges

The models developed for the *factor of frugality* seem simple but their scope is very wide. In particular, traditional and emerging areas of *additive manufacturing*, *machine learning*, *generative design*, *topology optimization* and *big data*, to name a few, can be employed with advantage in the service of F^S. Therefore, this will involve significant theoretical modeling and challenging experiments to accomplish the realization of an AFI. The interdisciplinary nature of *frugal engineering* will further add to the challenges of a *frugal* problem. However, the increasing availability of new technology and rapid progress in science and engineering will significantly aid in seeking planet-friendly solutions!

The challenges alluded to in the above paragraph warrant a talented workforce capable of looking for and subsequently solving *frugal* problems. Also, the systematic realization of frugal products, especially AFIs, warrants the employment of qualified workers with a good grasp of the fundamentals of science and engineering. The sophisticated nature of innovations listed by Rao (2013, 2017a) attests to the necessity of 'intelligent' designing of frugal innovations so that functionality is not compromised under a wide range of working conditions due to a no-frills streamlined structure. In fact, the development of these innovations should be grouped under research activities, and cutting-edge research should be employed to create myriads of frugal innovations. The sophistication inherent in these products has to be tackled through research that systematically applies science and other knowledge areas. The involvement of research is all the more important because frugality has something to offer to any sector of an economy. AFIs can appear in civil, mechanical, electrical, or any discipline of engineering; physical, chemical, or biological sciences; and even finance and economics. All that is required is a frugal mindset that is also adept in a given knowledge area. As far as engineering is concerned, research and development activities should be focused not least on the effective application of the frugal design approach; search for quantifying novel material-saving schemes; and employing cutting-edge areas such as artificial intelligence (AI) to strengthen the frugal approach due to its iterative nature showcased in Fig. 6.4 of Chap. 6.

13.5 Better Quality Control

This book has laid out the general principles of quality for frugal products. In particular, it has extended the *classical* principles of Juran to frugal engineering with lots of room for futuristic research into making this successful. The principles laid out in this book have to be fleshed out with details coming from a given frugal product/service in a given sector. This work has focused more on the quantitative aspects with a futuristic focus also on the people actually using the tools of quality control to deliver high-quality frugal products. In so doing, this book has provided an overview of the quality problem with frugal products with some areas needing futuristic efforts outlined in the following paragraphs.

The *supply chain* involved in any frugal product enterprise has to be coordinated to follow the quality principles necessary for producing frugal wares. The absence of suppliers' efforts in this regard would surely dent the expectations of the concerned maker of frugal products.

There is also a need to extend the ISO standards (ISO 1994) to the domain of frugal products. The extension of such a widely accepted standard will help immensely in adopting the principles of this book for frugal products. Moreover, devoting a portion of *total quality management* (TQM) (TQM 1997) to frugal products and, frugality in general, will along with ISO standards on frugality go a long way in affecting widespread adoption of frugal products. In fact, *classical* TQM stresses on the possibility of lowering costs by adhering to good quality (Blanton 1998) which can be used to advantage for frugality. Lastly, *awards* should be created to reward entities truly adhering to principles for producing successful frugal products. These rewards will go a long way in recognizing true frugal champions and also goad others into going frugal.

13.6 The Road Ahead

With our planet under the ever-tightening grip of *climate change* and other crises, the chances of adapting to a warming world are looking ever more probable. AFIs could significantly aid in the mass adoption of low-cost amenities to improve the living standards of one and all. In particular, air conditioning, transport and access to water are just the tip of the iceberg of utilities necessary for comfortable living in a warming world. The frugalization of as many of these amenities as possible through science, engineering and other knowledge areas will greatly help in our endeavors to have quality life in a warming world.

Since *frugal engineering* is a new discipline, there has to be concerted efforts on the part of researchers worldwide to get involved in it. The imposition of constraints on resource consumption will lead to novel ways of accomplishing existing technologies and also pave the way to discoveries in newfangled technologies. Therefore, large-scale targeted R&D efforts would significantly help in unraveling these technologies due to the involvement of rigorous science and other knowledge areas. And rigor in all manner of activities in *frugal engineering* is indispensable to its success and here again the involvement of researchers would facilitate the development of systematic methodologies.

The negative connotation associated with *Jugaad*, at least in North India, should be dispelled in light of the content covered in this book. *Advanced frugal innovations* are anything but "cheap" in their quality. The sophistication seen in AFIs has been built mostly through serious and rigorous application of science, engineering and relevant knowledge areas by workers with advanced skills or usually researchers. Their ingenious workings and robust quality performance attest to AFIs actually being a new class of products that show amazing advances made in science and technology through frugality. Even *grassroots frugal innovations* display the work of ingenuity by those clever and inventive craftsmen who have unwittingly hit on the right "science" for their products.

However, the existing literature is replete with efforts focused on the environmental and social credentials of FIs in general. This book entreats the *frugal innovation* community to start encouraging and/or making AFIs from scratch. This is because we can keep redefining the features of these products and their types but endeavoring into their making in different sectors would make a big difference to this world for the better. There are just two types of FIs, AFIs use engineering for their realization and GFIs are makeshift devices needing serious input of engineering to harness their true potential.

13.6.1 *Advanced Frugal Services (AFSs)!!*

The service sector should be revisited with a focus on advanced frugality. An *advanced frugal service* (AFS) might be proper not least, due to the proliferation of low-cost air travel. An AFS should be defined as delivering the best-quality service with minimal/suitable amount of resources at low cost. Here maintenance of resources should be of high quality due to significant use of limited resources.

13.6.2 *Fourth Industrial Revolution*

The title of this sub-section was coined by Schwab (2016) and it basically connotes a unifying trend that has blurred boundaries between physical and digital. The *fourth industrial revolution* includes the latest technologies such as *additive manufacturing*, *machine learning*, *synthetic biology*, *generative design* and everything else that exists and is yet to come through advances in both S&T and other knowledge areas. *Frugal engineering*, as presented in this book, is all encompassing in that it should ideally use advances in all knowledge areas to create mind-boggling AFIs. Moreover, the *factor of frugality* can be used as part of this new revolution as a modern *safety factor* to design and manufacture any product from scratch. It is natural that FE should be an integral part of this new revolution since humans are facing a bad combination of worst problems challenging their existence.

It is imperative that FE is embraced by all sections of society to maintain and continue uplifting our living standards and leave a planet in good shape for future generations. Therefore, the budding field of *frugal engineering* needs support not just from policymakers and governmental bodies, but also from society at large. The awareness in people for supporting products that are truly *frugal* while not sacrificing quality can go a long way in encouraging markets globally to root for AFIs. In this regard, the F^S can be used to discern products in various sectors for their *frugal* potential. The printing of F^S values on products in various sectors will also give consumers an awareness of whether undue resources have been wasted in producing them (Rao 2017b). Hopefully, such mass awareness of frugality in engineering and our way of life will help us tide over *climate change*.

And lastly, in this age of *climate change*, products should be referred to as *Advanced Frugal*-tech and not just *high*-tech! Because, *Advanced Frugal* is beyond *high*-tech since, as shown in this book, frugality blends a complex mixture of sophistication in engineering; lesser resources; and cost-effectiveness!

References

Blanton G (1998) Hidden costs to society. Qual Dig 18(6):18
IEA (2009) World energy outlook. Technical report, International Energy Agency (IEA)

IPCC (2001) Climate change 2001: the scientific basis. Contribution of working group i to the intergovernmental panel on climate change. Technical report, Intergovernmental Panel on Climate Change (IPCC)

ISO (1994) ISO 9000 handbook, 2nd edn. Irwin Professional Publishing, Fairfax, VA, USA

Liefner I, Losacker S, Rao B (2020) Scale up advanced frugal design principles. Nat Sustain

MacKay D (2009) Sustainable Energy—without the hot air. UIT Cambridge, UK

Rao B (2013) How disruptive is frugal? Technol Soc 35:65–73

Rao B (2017) Advances in science and technology through frugality. IEEE Eng Manag Rev 45(1):32–38

Rao B (2017) Revisiting classical design in engineering from a perspective of frugality. Heliyon 3(5):e00299

Rockström J, Steffen W, Noone K, Persson Å, Chapin FS, Lambin E, Lenton TM, Scheffer M, Folke C, Schellnhuber H, Nykvist B, Wit CAD, Hughes T, van der Leeuw S, Rodhe H, Sörlin S, Snyder PK, Costanza R, Svedin U, Falkenmark M, Karlberg L, Corell RW, Fabry VJ, Hansen J, Walker B, Liverman D, Richardson K, Crutzen P, Foley J (2009) Planetary boundaries: exploring the safe operating space for humanity. Ecol Soc 14(2):32

Schwab K (2016) The fourth industrial revolution. World Economic Forum

TQM (1997) A manifesto of TQM (1-4)—quest for a respectable organization presence. Societas Qualitas

SPRINGER NATURE

GPSR Compliance

The European Union's (EU) General Product Safety Regulation (GPSR) is a set of rules that requires consumer products to be safe and our obligations to ensure this.

If you have any concerns about our products, you can contact us on ProductSafety@springernature.com

In case Publisher is established outside the EU, the EU authorized representative is:

Springer Nature Customer Service Center GmbH
Europaplatz 3
69115 Heidelberg, Germany

The manufacturer's authorised representative in the EU is Springer Nature Customer Service Centre GmbH, Europaplatz 3, 69115 Heidelberg, Germany. If you have any concerns regarding our products, please contact ProductSafety@springernature.com

Printed and bound by CPI Group (UK) Ltd, Croydon, CR0 4YY
01/10/2026
02225902-0005